Elements of
Computational
Fluid Dynamics

ICP FLUID MECHANICS

ICP Fluid Mechanics
Vol. 2

Elements of Computational Fluid Dynamics

John D Ramshaw
Portland State University, USA

Imperial College Press

Published by

Imperial College Press
57 Shelton Street
Covent Garden
London WC2H 9HE

Distributed by

World Scientific Publishing Co. Pte. Ltd.
5 Toh Tuck Link, Singapore 596224
USA office: 27 Warren Street, Suite 401-402, Hackensack, NJ 07601
UK office: 57 Shelton Street, Covent Garden, London WC2H 9HE

British Library Cataloguing-in-Publication Data
A catalogue record for this book is available from the British Library.

ELEMENTS OF COMPUTATIONAL FLUID DYNAMICS
ICP Fluid Mechanics — Vol. 2

ISBN-13 978-1-84816-695-0
ISBN-10 1-84816-695-8
ISBN-13 978-1-84816-705-6 (pbk)
ISBN-10 1-84816-705-9 (pbk)

Typeset by Stallion Press
Email: enquiriesstallionpress.com

Printed in Singapore.

The lust for calculation must be tempered by periods of inaction...

— J. L. Synge

CONTENTS

PREFACE

The purpose of this book is to present the elements, i.e., the basic ingredients or building blocks, of computational fluid dynamics (CFD), from an admittedly personal point of view. In essence, it summarizes what I consider to be the ABC's of CFD. It is my firm conviction that the material covered herein constitutes the essential bare minimum of knowledge that should be possessed and clearly understood by all aspiring practitioners of CFD, both users of existing codes and methods, and *a fortiori* developers of new ones. I regard this knowledge as necessary to the intelligent practice of CFD, but it will certainly not be sufficient for more advanced students or researchers. The subject has become much too large to admit of a single encyclopedic treatment, but a list of more advanced books is provided in the Bibliography. This book does not intend to compete with more comprehensive treatments, but rather attempts to provide a detailed discussion of some of the more basic aspects of the subject that have not always received the emphasis they deserve. The discussion is entirely restricted to finite-difference methods, but many of the concepts discussed have close analogs in finite-element and spectral methods.

The main prerequisites assumed of the reader are a familiarity with the basic equations of fluid dynamics, some intuitive feeling for the physical meaning and mathematical character of the various terms therein, and a certain mathematical facility in manipulating the equations. Some previous exposure to numerical analysis would also be helpful, but is not essential and is not presumed. The required numerical analysis is either developed *ab initio* herein, or may readily be acquired from standard undergraduate textbooks or online sources.

CFD is the art and science of obtaining approximate numerical solutions to the equations of fluid dynamics. One might naively suppose that once the equations have been established by physical arguments, their solution could simply be regarded as a problem in mathematics to which their physical significance is no longer especially relevant. But nothing could be further from the truth–the physics and numerics are intimately intertwined and cannot be separated. In fact, many if not most numerical ailments possess a definite physical interpretation, an understanding of which is often invaluable to their proper diagnosis and treatment. This book accordingly has a strongly physical flavor, and relies heavily on arguments that most physicists and engineers will probably find satisfactory and convincing, but which are less likely to meet with the approval of mathematicians.

This book has not yet been used in the classroom, but various parenthetical details omitted from the main discussion have been relegated to numbered exercises. However, further problems and exercises would be highly desirable in a comprehensive CFD course, including simple programming exercises whereby students could obtain some hands-on familiarity and experience with the behavior of the various numerical methods discussed. The use of this book as the primary textbook for an introductory CFD course would therefore require some additional effort on the part of the instructor, so in its present form it may be more suitable as a supplementary or secondary reference for introductory courses in fluid dynamics and CFD. It might, however, be suitable for short courses of a single academic quarter or less. I also hope that it will provide an accessible introduction suitable for self-study by those who lack the time or opportunity to attend a formal course on the subject. I myself was in that situation at the beginning of my career, and I have tried to write the book that I wish had been available when I was first learning the subject in that way, as well as the textbook that I would use if I were teaching an introductory short course on CFD.

This book makes no pretense at historical scholarship, and with a single exception it does not refer to the research literature. Neither does it provide specific references to other books for more extensive or authoritative treatments of the particular topics discussed herein, or for further information about the various parenthetical topics that are mentioned in passing but are not discussed in any detail. Allusions of the latter type are numerous, and their purpose is simply to call the reader's attention to those topics and supply keywords to facilitate the search for more information. An abundance of additional information about topics of both types can readily be obtained from the books listed in the Bibliography, and by the use of online resources such as Wikipedia, Scholarpedia, Google Scholar, the ISI Web of Knowledge, and the DOE

OSTI Information Bridge, *inter alia*. The absence of specific references does not imply or suggest, and should not be construed as, any claim of originality for any of the concepts, ideas, techniques, methods, or results discussed in this book. Of course, this disclaimer does not apply to the manner in which the material has been organized and presented, for which the author bears sole responsibility.

Acknowledgments

I am indebted to C. H. Chang, L. D. Cloutman, R. W. Douglass, and G. A. Hansen for reviewing an earlier draft of the manuscript and offering many helpful comments and suggestions, which have immeasureably improved the final result. I am also grateful to the many colleagues with whom I have been privileged to interact or collaborate during the nearly forty years that I have been involved in CFD. They taught me much of what I know about the subject. The list is too long for me to mention everyone, but the following people deserve special acknowledgment: A. A. Amsden, T. D. Butler, C. H. Chang, L. D. Cloutman, J. K. Dukowicz, C. W. Hirt, J. A. Trapp, and last but not least, J. G. Trulio, who first introduced me to the subject, but I'm sure he meant well, and I forgive him.

I greatly appreciate and gratefully acknowledge the courtesy and kind hospitality extended to me by Professor Erik Bodegom and the Department of Physics at Portland State University. Finally, I thank Elsevier Ltd. for granting permission to use the epigraph on p. v, which was taken out of context from *Relativity: The General Theory*, by J. L. Synge (North-Holland, Amsterdam, 1966) but is profoundly pertinent to CFD as well.

RELATIONAL NOTATION

$\equiv$ means "is equal to by definition"

$\approx$ means "is approximately equal to"

$\sim$ means "is of the same order of magnitude as," and does *not* imply asymptotic equality

$\mathcal{O}(\varepsilon)$ denotes a term or terms of order ε; i.e., terms that become proportional to ε for sufficiently small ε, normally with a nonzero coefficient of proportionality

Chapter 1

INTRODUCTION

Fluid dynamics has many important practical applications, and is consequently of immense technological and economic significance. Unfortunately, the nonlinear partial differential equations of fluid dynamics can only be solved analytically in a few simple special cases. One must therefore resort to numerical methods to obtain solutions to most problems of practical interest. Computational fluid dynamics (CFD) is the art and science of computing such solutions. The subject encompasses both development and application of the numerical methods and computer codes used for that purpose.

1.1. The Structure of the Equations

The equations of fluid dynamics are time evolution equations of the general form

$$\frac{\partial \mathbf{v}}{\partial t} = \mathcal{D}(\mathbf{v}) \tag{1.1}$$

where $\mathbf{v} = \mathbf{v}(\mathbf{x}, t) = (v_1, v_2, \ldots)$ is a vector whose components are the dependent variables v_i such as the mass density ρ, velocity vector $\mathbf{u}$, and specific internal energy e of the fluid, all of which are functions of the position $\mathbf{x}$ and the time t, and $\mathcal{D}$ is a nonlinear vector differential operator that acts on the position-dependence of $\mathbf{v}$. In rectangular Cartesian spatial coordinates (x, y, z), which are sufficient for present purposes, $\mathbf{x} = x\mathbf{i} + y\mathbf{j} + z\mathbf{k}$, where $(\mathbf{i}, \mathbf{j}, \mathbf{k})$ are the constant unit vectors in the three coordinate directions. In general, $\mathbf{x}$ and $\mathbf{u}$ are three-dimensional vectors, but in problems with sufficient spatial symmetry they reduce to two-dimensional vectors or scalars, and the equations likewise reduce to two-dimensional or one-dimensional form.

Regardless of the detailed form of $\mathcal{D}(\mathbf{v})$, Eq. (1.1) governs the manner in which $\mathbf{v}(\mathbf{x}, t)$ evolves in time. Since the equations are first-order in time, they constitute an initial-value problem in which knowledge of $\mathbf{v}(\mathbf{x}, 0)$ at the initial time $t = 0$ uniquely determines the solution at any later time t (provided of course that the proper boundary conditions are also specified). Most fluid dynamics problems are of this "transient" or "unsteady" type, and they are accordingly solved numerically by means of time-marching methods that recursively advance the solution from one discrete time level to the next. This book is almost entirely concerned with methods of this type. However, there are also many situations in which steady-state solutions (i.e., solutions with $\partial \mathbf{v}/\partial t = \mathcal{D}(\mathbf{v}) = 0$) are of interest. The simplest (but not usually the most efficient) way of computing such solutions is as the asymptotic long-time limit ($t \to \infty$) of a transient calculation. In this way, time-marching methods can also be used to obtain steady solutions, and there are various ways of improving their efficiency for this purpose by artificially accelerating the approach to steady state. In contrast to transient solutions, however, steady-state solutions are not always unique (not even when the proper boundary conditions are imposed), but sometimes depend on the transient history of the flow from which they evolved. When this is (or is suspected to be) the case, the transient evolution should obviously not be artificially accelerated, as this could lead to the wrong solution. Conversely, stable steady-state solutions do not exist at all in many flow configurations, in which case an accurate transient calculation will reflect this behavior by remaining unsteady indefinitely (e.g., vortex shedding behind a circular cylinder).

The detailed form of the equations of fluid dynamics is defined by specifying the list of dependent variables v_i and the precise mathematical form of the operator $\mathcal{D}(\mathbf{v})$. A salient feature of fluid dynamics is that $\mathcal{D}(\mathbf{v})$ naturally separates into the sum of several terms that describe distinct physical processes, so that

$$\mathcal{D}(\mathbf{v}) = \sum_{\alpha} \mathcal{D}_{\alpha}(\mathbf{v}) \qquad (1.2)$$

in which the individual terms $\mathcal{D}_{\alpha}(\mathbf{v})$ represent such phenomena as advection, compression, pressure forces, viscous stresses, heat conduction, thermodynamic ($p\,dV$) work, diffusion, external forces, sources/sinks, friction, etc. The reader is presumed to be familiar with the usual mathematical expressions for such terms, which can be found in any fluid dynamics textbook. The particular terms considered in this book will be given in the individual chapters where they are discussed, and some common and familiar special cases of Eq. (1.1) can be found in Chapter 9.

The various terms $\mathcal{D}_\alpha(\mathbf{v})$ differ in mathematical character and are consequently susceptible to different numerical methods. Indeed, as we shall see, a numerical method that is well suited to one type of term may well be unconditionally unstable and hence entirely useless for another. Moreover, each of the individual terms $\mathcal{D}_\alpha(\mathbf{v})$ can be thought of as describing the entire dynamics of the fluid in the special case in which all of the other terms vanish. Thus the overall numerical method used for Eq. (1.1) in its entirety must be required to reduce, as a special case, to a viable numerical method for each of the more specialized equation systems

$$\frac{\partial \mathbf{v}}{\partial t} = \mathcal{D}_\alpha(\mathbf{v}) \tag{1.3}$$

These general considerations suggest that the behavior of both the equations themselves and the numerical methods used to solve them can best be understood by separately considering the various individual terms and their numerical approximations in isolation (or sometimes in pairs, as in the propagation of sound waves). The *leitmotiv* of this book is that the particular numerical expressions and equations used to approximate the individual terms $\mathcal{D}_\alpha(\mathbf{v})$, and the corresponding Eq. (1.3), constitute the elementary ingredients or basic building blocks of CFD. The numerical method for Eq. (1.1) as a whole is then assembled by combining those ingredients.

The synthetic approach outlined above is more than a mere expedient simplification, although it is that too, but is in fact essential to developing the insight and understanding necessary to intelligently construct a complete composite numerical method for solving Eq. (1.1) in its entirety. Conversely, one would intuitively expect that the behavior of the resulting composite method can largely be interpreted and understood in terms of its various elementary constituents. However, there is no rigorous mathematical justification for this expectation, and it is not inconceivable that interactions between those constituents will produce new and unexpected behavior, or even result in a composite method that is unsuitable for use in solving the full Eq. (1.1). Fortunately, a vast wealth of empirical evidence amply confirms that separate analyses of the individual ingredients in such composite methods do indeed suffice to account at least qualitatively, and more often than not semi-quantitatively (i.e., to within correction factors of order unity), for the vast majority of the observed behavior of the overall method.

1.2. The Form of the Equations

One might naively suppose that the next logical step would be to attempt to write out the most general possible form of the terms $\mathcal{D}_\alpha(\mathbf{v})$ in complete mathematical detail. For present purposes, however, this is unnecessary and would in fact be undesirable and counterproductive. For one thing, the essential features of those terms, and the numerical expressions used to approximate them, are easiest to discuss, analyze, and understand when they are presented in their simplest and most basic forms, not in full generality. The insight thus acquired then serves as a guide to subsequent generalizations and a foundation upon which they can be built. But there is a more fundamental reason why it would not be useful to attempt to write out the equations of fluid dynamics in full generality, namely that those equations are not in fact unique. The lack of uniqueness originates in two unrelated ways. First, the equations are not unique in the obvious sense that they can be mathematically transformed into a variety of equivalent forms, each of which is better suited to some types of problems and numerical methods than others. Second, the equations are inherently nonunique in a deeper and more subtle sense that may be surprising or even disconcerting to beginners: no completely general universally valid system of fluid dynamics equations even exists! Different variants of the equations are subject to different restrictions, employ different approximations and simplifications, and consequently apply to different classes of problems. Contrary to the impression fostered by certain introductory fluid dynamics texts, the familiar Navier–Stokes (N–S) equations for viscous compressible fluid flow were not carved by the gods on stone tablets. Among their many limitations is the restriction to Newtonian viscous stresses, which are diffusional in form and character and consequently predict the unphysical behavior that small disturbances propagate with infinite signal speed. (The resulting errors are normally negligible, but when they are not the equations require major modifications.) In their usual form, the N–S equations also neglect the effects of excited-state kinetic processes, such as vibrational nonequilibrium in chemical lasers or shock waves. For that matter, they do not provide a quantitatively accurate description of the internal structure of strong shock waves even in simple monatomic fluids, or of other phenomena occurring on atomic or molecular length scales. These remarks are not intended in any way to denigrate the venerable N–S equations, which represent a remarkably useful internally consistent physical theory and an impressive intellectual edifice. We merely wish to emphasize that those equations are not exact and should not be regarded as the "correct" equations of fluid dynamics.

Similar limitations are not peculiar to the Navier–Stokes equations. Any particular system of fluid dynamical (or more generally continuum-mechanical) equations is not exact but is merely a hopefully accurate approximation with a restricted domain of validity, and it will require modification to deal with problems in which its internal assumptions and approximations break down. The serious student of fluid dynamics should learn early on to embrace both the physical limitations of the equations and the corollary that there is no such thing as a single general universal system of equations that accurately describes the dynamics of fluids. It is consequently of the utmost importance to carefully consider and determine the most appropriate system of equations for any given fluid dynamics problem before making any attempt to solve them, either analytically or numerically. Incalculable resources of time, effort, and funding have been wasted in ill-conceived and ill-considered efforts to obtain accurate solutions to inappropriate, inaccurate, incomplete, and/or incorrect equations.

Of course, determining the appropriate equations to solve is a problem in physics rather than numerical analysis, and hence arguably lies outside the realm of CFD *per se*. However, such a narrow view of CFD, while perhaps logically defensible, is highly inadvisable. Modern computers are so powerful that they tend to foster a certain mental laziness; there is a temptation to "let the computer solve the problem," but the computer cannot do the thinking required to formulate the problem for solution. There is an all too common tendency among CFD practitioners to neglect or gloss over the physical aspects of the problem in favor of the computational ones, and to formulate the equations in a hasty or superficial way and rush headlong to their premature numerical solution. This temptation must be resisted at all costs; no matter how accurate the numerical method may be, it cannot coax a physically accurate solution from a physically inaccurate equation system. But conversely, no matter how accurate the equation system itself may be, the accuracy of its numerical solutions is limited by that of the numerical method used to compute them. Inaccurate solutions, whatever their cause, are not merely useless and wasteful, but are actively dangerous in situations where safety is at state (e.g., nuclear reactors). To paraphrase Mark Twain, it ain't what you don't know that hurts you — it's what you think you know that just ain't so.

Fortunately, although the equations of fluid dynamics are not, and cannot be expected to be, universal, essentially all such equations contain the same basic types of terms (advection, diffusion, etc.), and as a general rule each of those types is susceptible to similar numerical methods. Those methods are the elementary constituents or "elements" of the present treatment, much as carbon and hydrogen are the elementary constituents of the hydrocarbons, and they are much more nearly universal than the equations in which they appear. The primary purpose of this book is to familiarize the reader with those elements and their properties.

1.3. CFD Computer Codes

Before any particular CFD algorithm can be used to compute practical numerical solutions, it must be translated into a computer program. CFD programs are traditionally referred to as *codes*, and are usually written in a high-level programming language such as Fortran or C. The cautionary remarks of the previous section are equally applicable to CFD codes, and *a fortiori* to codes written by others, especially commercial CFD software. It is the responsibility of the project leader or principal investigator to ensure that he or she clearly understands and approves of both the equations being solved and the numerical methods that are being used to solve them. This includes convincing oneself, and being prepared to convince others, that the equations provide a sufficiently accurate physical description of the problem or process of interest, and that the numerical method is free from unacceptable limitations or approximations and can reasonably be expected to produce sufficiently accurate numerical solutions. The recursive nature of CFD calculations renders them susceptible to various instabilities and other pathologies. CFD codes consequently tend to lack robustness, and this limits the degree to which they can be routinely used as "black boxes" like most other software. There tends to be an inverse correlation between accuracy and robustness, so users of CFD codes should be aware that their developers may have been tempted to sacrifice some of the former to improve the latter. These caveats lead naturally to what is perhaps the eternal perennial question of CFD:

1.3.1. *What's in the code?*

The short answer to this question is that nobody knows, not even the author or authors. This pessimistic assessment is a slight exaggeration for effect, but it quite rightly emphasizes that it is by no means easy or straightforward to obtain a complete, definitive, or trustworthy answer to that critically important question. Indeed, "nobody knows" is uncomfortably close to the literal truth in the case of large and complex CFD codes, which are the rule rather than the exception. Unfortunately, the detailed merits and deficiencies of CFD codes cannot always be determined by consulting the formal code documentation. Most practical CFD codes are so large and complex that the documentation is highly unlikely to be either exhaustively complete or entirely accurate. In case of doubt, the only completely reliable documentation is the code itself, even if you yourself wrote both the code and the documentation. This presents an insoluble dilemma in the case of commercial CFD software that does not allow the user to examine the source code.

1.3.2. *Verification and validation*

Modern CFD codes often contain tens or even hundreds of thousands of lines of code. Writing such codes without errors obviously presents a formidable challenge, which is rarely rewarded by complete success. Even a hypothetical error-free code is only useful if the inherently approximate numerical solutions it produces are sufficiently accurate for the purposes at hand. Verification and validation ("V&V") of the codes is therefore an essential aspect of CFD. This process lies outside the scope of this book, but has become increasingly important (and more difficult) as rapid advances in computer technology have given rise to ever larger and more complex CFD codes. Nevertheless, V&V remains, as it always has been, a tedious but essentially straightforward and common-sense activity consisting primarily of obvious procedures like testing the individual components, modules, and subroutines in isolation, performing a wide variety of test calculations, comparing the results with known or manufactured analytical and previous numerical solutions, exercising all branches and logical pathways in the code, etc. While these and other useful guidelines can be given, it seems unlikely that V&V can ever be fully reduced to rigid, structured, or formalized prescriptions or procedures. Such procedures can become wasteful and counterproductive if they are instituted without appropriate checks and balances. Much like "quality,"

V&V in the abstract is something to which no rational person could object, and this makes it vulnerable to political exploitation. There is a further danger that otherwise reasonable V&V procedures may create a false sense of security if they are misinterpreted as providing a certified guarantee that each and every coding error has been detected and corrected, or that the code will invariably produce results with an accuracy that can be precisely quantified. Rigorous error bounds can sometimes be established in simple problems of purely academic interest, but such assurances cannot realistically be expected in complex practical engineering calculations with unknown exact solutions. The level of accuracy in real-world applications can rarely if ever be established with mathematical certainty, but can merely be estimated to a greater or lesser degree of plausibility that is inherently somewhat subjective.

1.3.3. *Visualization*

The use of suitable plotting and visualization software is another essential aspect of CFD praxis. CFD codes produce voluminous quantities of data, far too much for the human mind to directly comprehend. Restricting one's attention to a few quantities of primary interest (such as the total lift and drag on an airfoil) discards most of the insight that CFD calculations are capable of providing. Visualizing the numerical solution by means of appropriate plotting techniques can reveal at a glance previously unsuspected flow features of primary importance, suggest advantageous modifications in the geometrical configuration or operating conditions, or reveal "bugs" (errors) in the code that would be unlikely to be detected in any other way. Standard useful types of plots include contour and velocity vector plots, color level plots, surface and three-dimensional perspective plots, and various hybrids thereof. In the Good Old Days, which were not so good in this regard, it was customary for CFD code developers to write their own plotting routines and bundle them with the codes. This is fortunately no longer necessary, since various free and commercial plotting packages are readily available that are well suited for CFD applications. In short, practicing CFD without good visualization capabilities is like flying blind without instruments, and should not be attempted.

1.4. Organization of the Book

The numerical methods discussed in this book cannot be properly understood or analyzed without the use of some basic underlying concepts, tools, and techniques, including finite-difference approximations to derivatives, truncation-error analysis, and stability analysis. It is essential for the reader to become familiar and conversant with those preliminary topics before the structure and details of particular numerical methods can be fully appreciated, or even comprehended. Chapters 2–4 are accordingly devoted to a discussion of the necessary preliminary concepts and analytical tools. The individual numerical methods for the different types of terms occurring in CFD are then taken up in Chapters 5–8, which constitute the heart of the book. Finally, the manner in which those ingredients can be combined to construct complete CFD algorithms is discussed and illustrated in Chapter 9. This arrangement corresponds to the natural logical sequence of the material, but it has two drawbacks. First, it significantly delays the discussion of the particular numerical methods themselves, and this may be frustrating to readers who find the stockpiling of provisions tiresome and are impatient for the ship to set sail. Second, although the concepts and techniques discussed in Chapters 2–4 are not difficult, they are inherently general and somewhat abstract. They are consequently easier to grasp once their use has been illustrated by application to the specific examples considered in Chapters 5–8. Readers who find Chapters 2–4 unduly dense or tedious may therefore wish to skip ahead to Chapters 5–8, referring back to Chapters 2–4 as necessary.

Chapter 2

FINITE-DIFFERENCE APPROXIMATIONS

2.1. Discretization of Space and Time

Finite-difference approximations are defined with respect to sets of discrete spatial points and time levels sampled from the continuous spatial coordinates (x, y, z) and time t. The discrete spatial points are labeled by integer subscripts i, j, and/or k, while the discrete time levels are labeled by an integer superscript n. Thus x_i denotes the x-coordinate of the ith spatial point, and t^n denotes the nth time level (*not* the nth power of t!). In most of this book, we restrict attention for simplicity to the so-called *one-dimensional* case in which there is only a single spatial variable x. The spatial points define a *mesh* or *grid* that subdivides the spatial domain of interest, and the individual points are called *mesh points*, *grid points*, or *nodes* of the mesh. Advanced CFD methods often employ moving meshes and mesh points, which will not be considered here.

In most practical problems, the points x_i and t^n will not be equally spaced; i.e., the increments $x_{i+1} - x_i$ and $t^{n+1} - t^n$ between adjacent points are not constant but vary with i and n. This creates some unfortunate algebraic and notational complications, and also requires a considerable amount of interpolating and bookkeeping. Those complications are straightforward but tedious and require careful attention to detail, but they are rarely of much conceptual significance. The basic ideas of CFD (or anything else) are most easily and clearly discussed and assimilated in the absence of extraneous complications, so we shall assume equal spacing of mesh points and time levels throughout this book. The increments $x_{i+1} - x_i \equiv \Delta x$ and $t^{n+1} - t^n \equiv \Delta t$ are therefore constants, independent of i and n, so that $x_i = i\Delta x$ and $t^n = n\Delta t$. Meshes of this type are called *uniform.* The increments Δx and Δt are referred to as the *mesh spacing* and *time step*, respectively. The calculational process of advancing the numerical solution from one time level to the next, as discussed in Sect. 2.3 below, is also referred to as a time step, or alternatively a *cycle.* Thus, for example, we might say that some particular quantity needs to be reevaluated on every cycle or time step of the calculation.

The value of a dependent variable $f(x)$ at the point $x = x_i$ will be denoted by the more compact notation f_i; i.e., $f_i \equiv f(x_i)$. Similarly, $f^n \equiv f(t^n)$. In fluid dynamics we are usually concerned with dependent variables $f(x, t)$ that depend on both space and time, for which the obvious notation is $f_i^n \equiv f(x_i, t^n)$. The fluid dynamical variables satisfy partial differential equations involving partial derivatives with respect to both x and t. For simplicity, we shall temporarily suppress the time dependence and proceed as though x were the only independent variable. We shall, however, recognize the implicit dependence on t or n by writing spatial derivatives as partial derivatives.

The intervals or regions $x_i < x < x_{i+1}$ between adjacent mesh points are referred to as the *cells* or *zones* of the mesh. The mesh spacing Δx is consequently often referred to as the *cell size* or *zone size*. The cell centers or midpoints lie at the points $x_{i\pm 1/2} \equiv \frac{1}{2}(x_i + x_{i\pm 1}) = x_i \pm \frac{1}{2}\Delta x$. This notation is clumsy, so we shall often use the convenient (albeit unconventional) shorthand notation $i\pm \equiv i \pm \frac{1}{2}$, so that $x_{i\pm} = x_{i\pm 1/2}$. Similarly, the value of a dependent variable $f(x)$ at $x = x_{i\pm}$ will be denoted by $f_{i\pm} \equiv f(x_{i\pm})$. The mesh points x_i and cell centers $x_{i\pm}$ can be regarded as two overlapping interpenetrating meshes, which together are said to constitute a *staggered mesh.* In more than one space dimension, the mesh points define the *corners* or *vertices* of the cells, as well as the endpoints of the cell *edges.* In three dimensions, the cell edges also define the sides of the cell *faces*, while in two dimensions the cell edges, sides, and faces become synonymous.

Physical continuum theories, including fluid dynamics, are often formulated as coupled systems of equations involving more than one dependent variable. The mathematical structure of such systems is typically such that their numerical solution is facilitated by considering some of the dependent variables to be fundamentally located and computed at the mesh points x_i, while others are fundamentally located and computed at the cell centers x_{i+}. These two basic types of variables are referred to as *node centered* and *cell centered*, respectively. In many CFD methods, it is natural and traditional to consider thermodynamic variables such as density, pressure, and temperature to be cell centered, while the fluid velocity and fluxes (flow rates per unit area) of thermodynamic variables are node centered. (In multidimensional CFD, the analogous terms "vertex centered," "face centered," "edge centered," and "corner centered" are also used, with the obvious meanings.) The rationale for this conventional placement of variables involves considerations that have not yet been discussed but will become clear in due course, including the minimization of checkerboarding errors (which are discussed in the next section), and finite-volume methods for constructing finite-difference approximations (which will be discussed in Chapter 7). It is essential to always bear in mind the points at which each particular dependent variable f is fundamentally located and computed. Thus if f is node-centered then f_i simply refers to its computed value at the point x_i, whereas if f is cell-centered then its value f_i at $x = x_i$ is not directly computed but must be approximated in terms of its primary computed values $f_{i\pm}$. Such approximations are almost invariably required, and are normally done by averaging or interpolation. Their necessity is mainly due to the fact that some important composite variables are hybrids of both types, such as the mass flux and kinetic energy density, both of which involve products of cell-centered densities and node-centered velocities. For the most part, such approximations are straightforward but rather tedious, and they will be addressed at the appropriate places as necessary. But they should not be taken too lightly; they can be hazardous and require careful attention to detail, especially since differencing reduces the order of accuracy of approximations or interpolations, as will be discussed in Sect. 2.5.

As the term suggests, finite-difference methods approximate derivatives by finite differences. This might seem like a great leap backward to the dark ages before calculus, but such a leap is precisely what is needed in order to develop numerical approximations suitable for digital computers, which are discrete finite machines that are incapable of directly representing continuous functions or taking derivatives. We now proceed to show how such approximations can be constructed, and to discuss some of their more important features and properties.

2.2. Taylor Series and Truncation Errors

Finite-difference approximations to derivatives can be systematically derived by means of Taylor-series expansions. For example, let us expand

f_{i+1} in a Taylor series about the point $x = x_i$ to obtain

$$f_{i+1} = f_i + \Delta x \left(\frac{\partial f}{\partial x}\right)_i + \frac{1}{2}\Delta x^2 \left(\frac{\partial^2 f}{\partial x^2}\right)_i + \frac{1}{6}\Delta x^3 \left(\frac{\partial^3 f}{\partial x^3}\right)_i + \mathcal{O}(\Delta x^4)$$

(2.1)

and similarly

$$f_{i-1} = f_i - \Delta x \left(\frac{\partial f}{\partial x}\right)_i + \frac{1}{2}\Delta x^2 \left(\frac{\partial^2 f}{\partial x^2}\right)_i - \frac{1}{6}\Delta x^3 \left(\frac{\partial^3 f}{\partial x^3}\right)_i + \mathcal{O}(\Delta x^4)$$

(2.2)

It follows from Eq. (2.1) that

$$\left(\frac{\partial f}{\partial x}\right)_i = \frac{f_{i+1} - f_i}{\Delta x} + \mathcal{O}(\Delta x)$$

(2.3)

which implies that the finite-difference expression $(f_{i+1} - f_i)/\Delta x$ can be used to approximate $\partial f/\partial x$ at the point $x = x_i$, and that the error in doing so is of order Δx. This is called the *forward difference approximation* to $\partial f/\partial x$ at the point x_i. Similarly, it follows from Eq. (2.2) that

$$\left(\frac{\partial f}{\partial x}\right)_i = \frac{f_i - f_{i-1}}{\Delta x} + \mathcal{O}(\Delta x)$$

(2.4)

so that $\partial f/\partial x$ at the point $x = x_i$ can also be approximated by the *backward difference approximation* $(f_i - f_{i-1})/\Delta x$, again with an error of order Δx. Thus we see that *finite-difference approximations are not unique;* in fact there is an infinity of possible difference approximations to any particular differential expression. This freedom and ambiguity is both a blessing and a curse; it essentially provides one with the freedom to order from an infinite menu of entrees, of which some are inedible, many are mediocre at best, none is fully satisfactory, and the least unpalatable are very expensive. None of those approximations is mathematically equivalent to the original differential expression, and hence none of them preserves all of its properties and features (e.g., conservation, positivity, monotonicity, lack of upstream influence, dispersion relations, etc.). The construction of difference approximations is therefore a game of tradeoffs, wherein one decides which differential properties are most and least important and attempts to preserve the former, if necessary at the expense of the latter.

Notice that the forward difference approximation to $\partial f/\partial x$ at the point $x = x_{i-1}$ is the same as the backward difference approximation to $\partial f/\partial x$ at the point x_i. This illustrates that a given finite-difference *expression* does not by itself define a unique finite-difference *approximation*; one must also specify the point at which the approximation applies.

Equations (2.1) and (2.2) can be combined to obtain a more symmetrical *centered difference approximation* to $\partial f/\partial x$ at the point x. Subtracting Eq. (2.2) from Eq. (2.1), we obtain

$$\left(\frac{\partial f}{\partial x}\right)_i = \frac{f_{i+1} - f_{i-1}}{2\Delta x} + \mathcal{O}(\Delta x^2) \tag{2.5}$$

which shows that the error in approximating $(\partial f/\partial x)_i$ by the centered difference expression $(f_{i+1} - f_{i-1})/(2\Delta x)$ is of order Δx^2, and will therefore normally be much smaller than the previous errors of order Δx, at least when Δx is sufficiently small. The power of Δx to which the error of a difference approximation is proportional is referred to as the *order* of the approximation. Thus we say that the forward and backward difference approximations to $\partial f/\partial x$ are first-order accurate, while the centered difference approximation is second-order accurate. The errors in such approximations are referred to as *truncation errors*, since they are incurred by truncating the Taylor series after a finite number of terms. They are also often referred to as *discretization errors*, since they result from the discretization of space and time. We shall regard those terms as synonymous and use them interchangeably, but the reader is cautioned that some authors draw a distinction between them and reserve the term "discretization errors" to refer to errors in the numerical solution rather than in the terms and equations used to compute it. Those two types of errors are distinct but closely related, as will be discussed in Sect. 3.1.

Unfortunately, the higher accuracy of the centered difference approximation comes with a price: it has the disadvantage that the value of f_i does not contribute to the approximate $\partial f/\partial x$ at the same point x_i. In many situations this lack of coupling allows the gradual accumulation of small errors, so that the numerical solutions at even and odd mesh points slowly drift away from each other. This effect is sometimes referred to as *checkerboarding*, and typically manifests itself as parasitic artificial oscillations with a wavelength of $2\Delta x$; i.e., "every-other-cell" oscillations. The essence of such oscillations is that a function of the form $f_i = f_0(-1)^i$ makes a zero contribution to Eq. (2.5), regardless of how large f_0 is. Various remedies for this problem have been developed, including a variety of algorithms generically referred to as *node couplers*, but such methods lie outside the scope of an elementary treatment. Of course, the simplest remedy is to use difference approximations that are not vulnerable to this problem. A primary advantage of staggered meshes is that they greatly facilitate the construction of such approximations, but not always with complete success.

Finite-difference approximations to $\partial f/\partial x$ at the cell centers x_{i+} can be developed in an entirely similar manner, and we readily obtain

$$\left(\frac{\partial f}{\partial x}\right)_{i+} = 2\left(\frac{f_{i+1}-f_{i+}}{\Delta x}\right) + \mathcal{O}(\Delta x) \tag{2.6}$$

$$\left(\frac{\partial f}{\partial x}\right)_{i+} = 2\left(\frac{f_{i+}-f_i}{\Delta x}\right) + \mathcal{O}(\Delta x) \tag{2.7}$$

$$\left(\frac{\partial f}{\partial x}\right)_{i+} = \frac{f_{i+1}-f_i}{\Delta x} + \mathcal{O}(\Delta x^2) \tag{2.8}$$

which respectively provide forward, backward, and centered difference approximations to $(\partial f/\partial x)_{i+}$. Note, however, that if the primary computed variables are the f_i, then the values of f_{i+} are not known or available *a priori*, so that the only one of Eqs. (2.6)–(2.8) that can be directly used without further interpolation or approximation is Eq. (2.8), which fortunately is also the most accurate. Comparing Eqs. (2.8) and (2.3), we see that the finite-difference expression $(f_{i+1}-f_i)/\Delta x$ is simultaneously both a first-order accurate forward difference approximation to $\partial f/\partial x$ at $x = x_i$, and a second-order accurate centered difference approximation to $\partial f/\partial x$ at $x = x_{i+}$. It is important to understand that those two approximations are conceptually different: they approximate $\partial f/\partial x$ at different values of x, and they consequently also differ in accuracy as shown by their truncation errors.

Exercise 2.1. Show that the expression $(f_{i+1}-f_i)/\Delta x$ is actually a first-order accurate approximation to $\partial f/\partial x$ for any $x \in [x_i, x_{i+1}]$, or for that matter any x such that $|x - x_{i+}| \sim \Delta x$.

Equations (2.1) and (2.2) can also be combined to obtain a finite-difference approximation to the second derivative $\partial^2 f/\partial x^2$ at the point $x = x_i$. Adding Eqs. (2.1) and (2.2), we obtain

$$\left(\frac{\partial^2 f}{\partial x^2}\right)_i = \frac{f_{i+1}-2f_i+f_{i-1}}{\Delta x^2} + \mathcal{O}(\Delta x^2) \tag{2.9}$$

which is a very common standard form that should be committed to memory.

2.3. Time Differencing

The above discussion has focused on spatial derivatives and differences. Time differencing is entirely analogous, but in the present context has some peculiar features of its own. The first such feature is that in this book we shall have no occasion to deal with time derivatives higher than the first, so we may restrict attention to finite-difference approximations to $\partial f/\partial t$. Just as for spatial derivatives, we may define forward and backward difference approximations to $\partial f/\partial t$, the derivation and form of which are entirely analogous to Eqs. (2.3) and (2.4):

$$\left(\frac{\partial f}{\partial t}\right)^n = \frac{f^{n+1} - f^n}{\Delta t} + \mathcal{O}(\Delta t) \tag{2.10}$$

$$\left(\frac{\partial f}{\partial t}\right)^n = \frac{f^n - f^{n-1}}{\Delta t} + \mathcal{O}(\Delta t) \tag{2.11}$$

However, it is convenient to shift the time level in Eq. (2.11) so that the finite-difference approximation to the derivative always involves the time levels n and $n+1$. We therefore rewrite Eq. (2.11) in the equivalent form

$$\left(\frac{\partial f}{\partial t}\right)^{n+1} = \frac{f^{n+1} - f^n}{\Delta t} + \mathcal{O}(\Delta t) \tag{2.12}$$

The reason this is done is that we will be using such approximations within the context of initial-value problems, in which numerical solutions are generated by means of *time-marching* (or time-stepping) methods which recursively advance the solution from one time level to the next. The initial conditions at $t = 0$ define the dependent variables at $n = 0$, from which their values at $n = 1$ can be computed. Their values at $n = 2$ can then be computed, in precisely the same way, from those at $n = 1$, and so on. It is therefore convenient to always denote the two time levels that are thus connected by n and $n+1$, so that n represents the time level at which the solution has just been computed and is therefore now known, while $n+1$ represents the succeeding time level at which the solution must be computed next. Time level n is often referred to as the *current, previous,* or *old* time level, while $n+1$ is referred to as the *advanced, next,* or *new* time level. The process of advancing the numerical solution from the current time level to the next one is often referred to as the *current time step* of the calculation.

For the reason just discussed, the time-centered finite-difference approximation to $\partial f/\partial t$ will be defined with reference to time levels n

and $n + 1$ rather than $n - 1$ and $n + 1$, and is therefore entirely analogous to Eq. (2.8):

$$\left(\frac{\partial f}{\partial t}\right)^{n+} = \frac{f^{n+1} - f^n}{\Delta t} + \mathcal{O}(\Delta t^2) \tag{2.13}$$

where the superscript $n+$ is an abbreviation for $n + 1/2$, and indicates that the quantity to which it is affixed is evaluated at time $t = t^n + \frac{1}{2}\Delta t = (n + \frac{1}{2})\Delta t$. In summary, the same finite-difference expression $(f^{n+1} - f^n)/\Delta t$ serves as a forward difference approximation to $(\partial f/\partial t)^n$, a backward difference approximation to $(\partial f/\partial t)^{n+1}$, and a centered difference approximation to $(\partial f/\partial t)^{n+}$, and which of them it represents must be clearly stated in order to unambiguously define the approximation.

2.4. Truncation-Error Analysis

Not all finite-difference approximations are explicitly constructed using Taylor series. Indeed, it is more common for experienced numerical analysts to construct such approximations by inspection and intuition, based on insights obtained from their previous experience with similar differential expressions and equations. A primary objective of this book is to assist the reader in developing this facility, so that when certain common types of terms are encountered, the various difference approximations that are likely to be suitable or unsuitable will immediately leap to mind. But regardless of how a particular approximation is constructed or from whence it originated, its accuracy should always be assessed or confirmed by means of a truncation-error analysis based on Taylor series expansions of the type employed above. Such analyses determine the order of accuracy of the approximation for small Δx and Δt. This information is especially critical in the case of improper approximations for which the truncation errors do not vanish but remain finite as Δx and/or Δt approach zero. Such approximations are called *inconsistent* and their accuracy is formally of zeroth order. They are not always obvious on inspection; approximations that superficially look very reasonable or natural *a priori* sometimes turn out to be inconsistent. As a general rule, such approximations are unacceptable and should be avoided at all costs, but there are occasional exceptions as discussed below.

Order of accuracy, as ascertained by truncation-error analysis, is such a common and useful way of characterizing and classifying finite-difference approximations that its limitations are not always sufficiently appreciated.

There is a tendency to assume that higher-order approximations are in general more accurate than, and consequently are always preferable to, lower-order approximations. Indeed, this view has been elevated to the status of dogma in some quarters of the CFD community. However, that assumption is justified only when the increments Δx and/or Δt are sufficiently small that the Taylor series expansions converge rapidly; i.e., within the first few terms. This is a very stringent condition that is not always satisfied in practical calculations, and when it is violated the situation is frequently reversed; for relatively large Δx and/or Δt, lower-order approximations can and often do produce more accurate results, both qualitatively and quantitatively, than higher-order approximations. We shall encounter several examples of this phenomenon later in this book. Cases are even known in which inconsistent approximations produce more accurate results for fixed finite values of Δx and/or Δt than standard consistent approximations do. The moral of the story is that although the formal order of accuracy and behavior of an approximation for small Δx and/or Δt are extremely important, they are not the only considerations. In practical situations, it is also important to ensure that the approximation still exhibits qualitatively reasonable behavior even when Δx and/or Δt are not small enough for the Taylor series to converge rapidly.

There are other tradeoffs to be considered in the decision of whether to employ higher- vs. lower-order finite-difference approximations, and most of them favor the latter. In particular, higher-order approximations are in general more complicated than lower-order ones, and they are consequently more difficult and costly to develop, program, debug, execute, and modify. In view of this, it is well to keep in mind that whatever additional accuracy higher-order approximations provide can in principle also be obtained from lower-order approximations simply by further reducing Δx and Δt in the latter. It should also be noted that in problems with legitimate physical discontinuities, such as shock waves, the formal order of accuracy of the approximation for continuous functions becomes entirely irrelevant, at least in the vicinity of the discontinuity. In such problems it can be shown that the maximum accuracy attainable in the solution itself is typically limited to second order, regardless of the formal order of accuracy of the finite-difference approximation as determined by a Taylor series expansion. This limitation is hardly surprising, since discontinuous functions cannot be expanded in Taylor series. Many numerical methods approximate such discontinuities as rapid but continuous transition regions spread over a few finite-difference cells or time steps. However, such techniques do not remove the above limitation, because the Taylor series do not converge rapidly (if at all) in such regions, so that higher-order truncation errors cannot be presumed to be small compared to lower-order ones. The reason is simply that the length and time scales over which the dependent variables change significantly are then inherently of the same order as Δx and Δt, so the latter increments are not small compared to those scales as required for rapid convergence of the Taylor series. Moreover, the individual truncation-error terms of all orders then diverge as Δx and Δt tend to zero, and hence provide no useful information about the actual truncation errors in that limit.

2.5. Approximations within Approximations

This is one of the more important sections in the entire book, because it discusses some very common and highly insidious hazards and pitfalls that have enticed even experienced numerical analysts into serious errors. The essential *leitmotiv* of the discussion is that *differencing reduces the order of accuracy*. What this vague statement actually means is best clarified by specific examples, but we must first establish the context. In the preceding development, it has been tacitly presumed that the true value of a typical dependent variable $f(x)$ is known at the points $x = x_i$, but only at those points. Most of the finite-difference approximations that we have considered have accordingly been expressed in terms of the quantities $f_i = f(x_i)$, and the analysis of their errors has been based on the above presumption. We have also remarked that finite-difference expressions often involve the values of the various dependent variables at points where they are not fundamentally located, and that those values must then be obtained by further approximation or interpolation. Thus, for example, Eq. (2.6) involves f_{i+}, and consequently cannot be used to approximate $(\partial f/\partial x)_{i+}$ until f_{i+} has itself first been independently approximated in terms of the values of the f_i. In situations of this type, one is perforce required to make *approximations within approximations* in order to proceed, and this has serious implications as to the overall accuracy of the resulting composite approximation. Those implications have not always been sufficiently appreciated or even recognized, and we now proceed to discuss them in more detail.

Continuing with the example of Eq. (2.6), let us suppose that f_{i+} has somehow been approximated in terms of the quantities f_i by an expression

$\tilde{f}_{i+}$ which is kth-order accurate; i.e.,

$$f_{i+} = \tilde{f}_{i+} + \mathcal{O}(\Delta x^k) \tag{2.14}$$

Combining Eqs. (2.6) and (2.14), we obtain

$$\left(\frac{\partial f}{\partial x}\right)_{i+} = 2\left(\frac{f_{i+1} - \tilde{f}_{i+}}{\Delta x}\right) + \mathcal{O}(\Delta x) + \mathcal{O}(\Delta x^{k-1}) \tag{2.15}$$

Now consider the special case $k = 1$, in which the approximation of f_{i+} by $\tilde{f}_{i+}$ is first-order accurate and is therefore still consistent. In that case, however, Eq. (2.15) shows that the resulting finite-difference approximation to $(\partial f/\partial x)_{i+}$ is zeroth-order accurate; i.e., inconsistent, and hence would normally be considered unacceptable. Thus we see that *consistently approximating a quantity that appears within another consistent approximation does not imply that the resulting overall approximation is consistent.* The importance of this observation can hardly be overstated, since many if not most people seem to find it counterintuitive at first, and failure to understand it has led to many errors, some of which have persisted in the literature and/or common practice for decades.

Inspection of the preceding example shows that the difficulty arises because finite-difference approximations to first derivatives involve dividing by Δx, so the order of accuracy of the quantity being differenced is effectively reduced by one. Similarly, approximations to second derivatives, such as Eq. (2.9), involve dividing by Δx^2, so that the order of accuracy of the quantities in the numerator is effectively reduced by two, and so on. Exceptional cases sometimes occur in which the lowest-order errors in the final result cancel out, thereby restoring the consistency. However, it would obviously be foolhardy to employ formally inconsistent approximations in the hope that their accuracy will ultimately be increased to first order by a fortuitous cancellation of zeroth-order errors. In any case, such cancellations rarely occur in practical meshes where $x_{i+1} - x_i$ varies with i.

The above points are so important that we shall proceed to belabor them by means of further examples, in accordance with the principle that no harm is done by beating a dead horse, which after all is hardly in any danger of dying again. As a second example, we consider the problem of constructing finite-difference approximations to second derivatives near physical boundaries. We shall suppose that there is a physical boundary at the point $x_1 + \frac{1}{2}\Delta x = x_{1+}$, and that the computational region is defined by $x \geq x_{1+}$. Setting $i = 2$ in Eq. (2.9), we obtain

$$\left(\frac{\partial^2 f}{\partial x^2}\right)_2 = \frac{f_3 - 2f_2 + f_1}{\Delta x^2} + \mathcal{O}(\Delta x^2) \tag{2.16}$$

However, the point x_1 lies a distance $\frac{1}{2}\Delta x$ outside the computational region, so the value of f_1 is not computed and is not known *a priori*, whereas f_2 and f_3 are primary computed variables. In this situation, the value of f_1 is in fact artificial and is available for use in setting the boundary condition, which we presume is a Dirichlet condition that prescribes the value of $f(x_{1+}) = f_{1+}$. The artificial value of f_1 must then be approximated in terms of f_{1+}, f_2, and/or f_3 so that Eq. (2.16) can be used to approximate $(\partial^2 f/\partial x^2)_2$. A simple, obvious, and superficially reasonable approximation of this type, and one that has been widely employed, is to obtain the value of f_1 by linearly extrapolating the values of f_2 and f_{1+}, so that $f_1 = 2f_{1+} - f_2$. This is equivalent to regarding f_{1+} as the arithmetical average of f_1 and f_2, which again seems eminently reasonable. Combining this approximation with Eq. (2.16), we obtain

$$\left(\frac{\partial^2 f}{\partial x^2}\right)_2 \approx \frac{f_3 - 3f_2 + 2f_{1+}}{\Delta x^2} \tag{2.17}$$

However, expanding f_3 and f_{1+} in Taylor series about the point x_2, we readily find that

$$\frac{f_3 - 3f_2 + 2f_{1+}}{\Delta x^2} = \frac{3}{4}\left(\frac{\partial^2 f}{\partial x^2}\right)_2 + \mathcal{O}(\Delta x) \tag{2.18}$$

so that Eq. (2.17) is in fact inconsistent and only accounts for $3/4$ of the value of $(\partial^2 f/\partial x^2)_2$. The problem is that the approximation $f_1 = 2f_{1+} - f_2$ is only second-order accurate, as is easily confirmed by another Taylor series analysis, and hence is effectively degraded to zeroth-order accuracy when it is divided by Δx^2 as required by Eq. (2.16). Physical boundary-value problems to which this example is directly relevant include fixed-temperature walls in heat conduction, and the no-slip boundary condition on the tangential velocity at solid walls in viscous flow.

Exercise 2.2. Construct a third-order accurate approximation for f_1 in terms of f_{1+}, f_2, and f_3 so that Eq. (2.16) provides a consistent first-order accurate approximation to $(\partial^2 f/\partial x^2)_2$. (*Hint:* multiply Eq. (2.18) by $4/3$, combine the result with Eq. (2.16), and solve for f_1.)

The aforementioned horse protesteth not its treatment, and *qui tacet consentit*, so we proceed to discuss a final example involving a dependent variable $f = f(x,t)$ that depends on both position and time. Let us consider the use of Eq. (2.9) to approximate $\partial^2 f/\partial x^2$ at $x = x_i$ and $t = t^{n+1}$:

$$\left(\frac{\partial^2 f}{\partial x^2}\right)_i^{n+1} = \frac{f_{i+1}^{n+1} - 2f_i^{n+1} + f_{i-1}^{n+1}}{\Delta x^2} + \mathcal{O}(\Delta x^2) \tag{2.19}$$

This approximation occurs in fully implicit time-marching numerical methods for diffusional processes; e.g., Eq. (6.14) in Sect. 6.2. As will be discussed in Sect. 3.3, the presence of the quantities $f_{i\pm1}^{n+1}$ in such methods requires the solution of a large sparse system of linear equations to determine f_i^{n+1} for all i and thereby advance the solution from time level n to time level $n + 1$. Novice numerical analysts often experience the illusory epiphany that this complication could be avoided by the simple expedient of approximating $f_{i\pm1}^{n+1}$ in Eq. (2.19) by $f_{i\pm1}^n$. This approximation might seem superficially reasonable on the grounds that Q^n is a consistent first-order accurate approximation to Q^{n+1}, or *vice versa*, for any Q. If we were to replace $f_{i\pm1}^{n+1}$ by $f_{i\pm1}^n$ in Eq. (2.19) on that basis, we would thereby obtain the approximation

$$\left(\frac{\partial^2 f}{\partial x^2}\right)_i^{n+1} \approx \frac{f_{i+1}^n - 2f_i^{n+1} + f_{i-1}^n}{\Delta x^2} \tag{2.20}$$

However, expanding f_{i+1}^n and f_{i-1}^n in Taylor series about the point $(x,t) = (x_i, t^{n+1})$, we readily obtain

$$\frac{f_{i+1}^n - 2f_i^{n+1} + f_{i-1}^n}{\Delta x^2} = \left(\frac{\partial^2 f}{\partial x^2}\right)_i^{n+1} + \mathcal{O}\left(\frac{\Delta t}{\Delta x^2}\right) \tag{2.21}$$

This shows that the approximation of Eq. (2.20) is inconsistent in the sense that its error does not necessarily vanish as Δt and Δx independently approach zero. The error does become negligible for sufficiently small $\Delta t/\Delta x^2$, but that condition requires impractically small values of Δt. This example illustrates the useful rule of thumb that the terms within finite-difference approximations to spatial derivatives should not be evaluated at different time levels, which normally results in an inconsistent approximation. Another textbook illustration of the same type of inconsistency is provided by the well-known Du Fort–Frankel method. The basic origin of such inconsistencies is again simply the presence of Δx in the denominator, which effectively reduces the order of accuracy of any approximations made in the numerator.

Exercise 2.3. Determine the coefficient of $\Delta t/\Delta x^2$ in Eq. (2.21).

Chapter 3

FINITE-DIFFERENCE EQUATIONS

Just as a differential equation is an equation involving derivatives, a finite-difference equation is an equation involving finite differences. The calculus of finite differences and difference equations is a useful and independent subject in its own right, but in the present context our interest in such equations is limited to their use as approximations to ordinary or partial differential equations. Such finite-difference equations are usually constructed by approximating the derivatives in the differential equations in terms of finite-difference expressions of the type discussed in the preceding chapter. As a simple example, which will be discussed in detail in Chapter 6, approximating the space and time derivatives in the diffusion equation (6.1) by means of Eqs. (2.9) and (2.10) results in the finite-difference equation (6.6). The resulting difference equations are variously referred to as finite-difference numerical methods, schemes, algorithms, analogs, or approximations to the original differential equations, and can then be used to compute approximate numerical solutions to the latter. In CFD, we are primarily concerned with time-marching schemes in which the finite-difference equations are used to advance the numerical solution from the old time level n to the new time level $n + 1$. Numerical schemes based on Eqs. (2.10), (2.12), and (2.13) involve only those two time levels and are accordingly called *two-level schemes*. We shall not consider multilevel schemes involving three or more time levels, the simplest of which are the three-level *leapfrog* schemes in which $(\partial f/\partial t)^n$ is typically approximated by $(f^{n+1} - f^{n-1})/(2\Delta t)$. We shall also not discuss two- or multi-step schemes (e.g., predictor-corrector methods), time-splitting schemes, or fractional-step methods, in which the numerical solution is advanced from time level

n to time level $n + 1$ in a sequence of intermediate steps. The reader should note, however, that such schemes are of considerable practical importance, particularly for numerical simulations of compressible flow.

3.1. Accuracy Considerations

The accuracy of the approximate solutions computed by finite-difference methods is obviously of paramount importance. Unfortunately, it is very difficult, and in most cases practically impossible, to precisely quantify such solution errors. The reason is that the solution error at any given space-time point (x, t) is the difference between the approximate and exact solutions at that point, and the latter is of course not normally known. In this situation, the best that can be done is to attempt to quantify the accuracy with which the difference scheme approximates the differential equation itself rather than its solution. The standard procedure for doing so is to determine the order of accuracy of the scheme as a whole by means of a truncation-error analysis based on Taylor series expansions. This is done in the obvious way simply by performing truncation-error analyses for each of the terms in the finite-difference equation, as described in the preceding chapter, combining them, and comparing the final result with the original differential equation. It is of course essential for this purpose to expand each term in a particular equation about the same space-time point. The manner in which such analyses are performed in practice will be illustrated in the examples considered in later chapters. Finite-difference schemes can thereby be classified as first-order, second-order, etc. in each of the independent variables. For example, many common simple schemes are first order in time and second order in space.

Numerical schemes for which the solution error tends to zero in the limit of vanishing Δx and Δt are said to be *convergent*. For sufficiently smooth solutions, the numerical solution and the difference scheme are normally expected to be of the same order of accuracy, so that the truncation errors of the latter should provide an indication of the accuracy of the former. More precisely, the *convergence rate* at which the solution errors tend to zero is ordinarily the same as the rate at which the truncation errors in the finite-difference scheme tend to zero. Even so, this behavior does not immediately provide a quantitative numerical estimate of the absolute solution error for fixed finite values of Δx and Δt. However, such estimates can be constructed by extrapolation procedures such as *Richardson extrapolation*. For example, if a numerical scheme is kth-order accurate in space, then for sufficiently small values of Δx one would expect the numerical solution $\tilde{f}$ to differ from the exact solution f by terms of order Δx^k; i.e.,

$$f \approx \tilde{f} + A \, \Delta x^k \tag{3.1}$$

where A is an as yet unknown constant. The value of A can be determined by computing the solutions $\tilde{f}_a$ and $\tilde{f}_b$ for two different spatial increments Δx_a and Δx_b, so that

$$f \approx \tilde{f}_a + A \, \Delta x_a^k \tag{3.2}$$

$$f \approx \tilde{f}_b + A \, \Delta x_b^k \tag{3.3}$$

Subtracting these two approximations and solving for A, we obtain $A \approx -(\tilde{f}_b - \tilde{f}_a)/(\Delta x_b^k - \Delta x_a^k)$, which combines with either Eq. (3.2) or (3.3) to provide an estimate of the true solution f and hence the accuracy of the approximate numerical solutions $\tilde{f}_a$ and $\tilde{f}_b$.

The order of accuracy of the numerical scheme is also useful in "debugging" (i.e., detecting and correcting errors in) the computer code used to compute the numerical solutions. This can be done by verifying that the solution errors tend to zero at the rate predicted by relations such as Eq. (3.1). For test problems in which the exact solution f is known, it suffices to perform the calculation with two different resolutions as discussed above. The first calculation is used to determine the coefficient A by means of Eq. (3.2), and one can then ascertain whether or not the second calculation satisfies Eq. (3.3) with approximately the same value of A. For problems in which the exact solution is not known, approximate values of both f and A must first be determined from Eqs. (3.2) and (3.3). A third calculation with a third value of Δx is then performed so that one can ascertain whether the resulting numerical solution $\tilde{f}$ satisfies Eq. (3.1) with approximately the same values of f and A. Of course, tests of this type are only meaningful when all the values of Δx involved are sufficiently small, and ensuring this may require some numerical experimentation.

3.2. Truncation Errors and Modified Equations

So far we have primarily focused on truncation-error analysis as a useful tool for determining the consistency and order of accuracy of finite-difference approximations and equations. For this purpose it suffices to examine the *order* of the truncation errors. However, the mathematical *form* of the truncation errors, in particular the derivatives appearing therein, provides a great deal of additional information and valuable insight into the physical and mathematical character of the approximation, and this insight often allows one to anticipate and physically interpret the qualitative behavior of difference equations and their solutions. In particular, such analyses can reveal the numerical stability or instability of the finite-difference equations (which is the subject of Chapter 4), as well as the qualitative character of the solution errors. The basis for these statements is the observation that the truncation errors effectively modify the form of the original differential equations, and it is those modified differential equations that the difference scheme is effectively solving. That is to say, the truncation-error terms represent the difference between the true differential equations and the finite-difference equations that are actually being solved. By their very nature, those terms are *differential* in form, and as such they sometimes possess a clear physical interpretation. For example, truncation errors that are diffusional in form provide insight into both numerical instability and artificial diffusion (or "numerical viscosity"), as will be discussed in Chapter 7.

The concept of the modified differential equation to which a particular finite-difference equation is equivalent can be made more precise as follows. Consider the differential equation $\mathcal{L}(f) = 0$, where $\mathcal{L}$ is a particular differential operator, and let $\mathcal{L}_\Delta$ be a particular finite-difference approximation to $\mathcal{L}$. The finite-difference equation $\mathcal{L}_\Delta(f_\Delta) = 0$ is then an approximation to the true differential equation, and its exact solution f_Δ is a numerical approximation to the true solution f. If the operators $\mathcal{L}$ and $\mathcal{L}_\Delta$ are both applied to the same continuous test function g, the truncation error of the finite-difference approximation is given by $\mathcal{E}(g) \equiv \mathcal{L}(g) - \mathcal{L}_\Delta(g)$. By means of Taylor series expansions of the type employed in the previous chapter, it is clear that $\mathcal{L}_\Delta(g)$ and $\mathcal{E}(g)$ can be expressed as infinite series involving derivatives of g of all orders. Once this has been done, the difference equation $\mathcal{L}_\Delta(f_\Delta) = 0$ can be reinterpreted as a differential equation which determines a smooth function f_Δ that coincides with the numerical solution at the discrete points $(x, t) = (x_i, t^n)$ where the latter is defined. It is then natural to regard that differential equation as a *modified equation* which corresponds to and approximates the original differential equation. Such modified equations are normally written in the equivalent form $\mathcal{L}(f_\Delta) = \mathcal{E}(f_\Delta)$ to compare and contrast their form with that of the original differential equation $\mathcal{L}(f) = 0$. They are formally differential equations of infinite order, since they contain derivatives of all orders, but the higher derivatives are multiplied by higher powers of the small increments Δx and Δt, so the higher-order terms can be presumed negligible for sufficiently smooth functions.

We have been careful to distinguish notationally between the exact analytical solution f and the approximate numerical solution f_Δ. However, in formulating and analyzing difference schemes this notational distinction is normally suppressed for simplicity, and it is simply understood that the dependent variables f appearing in the difference equations represent the approximate numerical solution rather than the unknown true solution. This is so obvious that there is a tendency to lose sight of it, and it must be kept in mind when manipulating the truncation error terms appearing in the modified equations, which as shown above involve derivatives of f_Δ rather than f. Such manipulations frequently involve transforming time derivatives into space derivatives, and this must be done using the modified equation rather than the original differential equation, since f_Δ does not of course satisfy the latter.

The above general considerations are easier to grasp within the context of specific examples. The most familiar, and probably the most important, examples are the diffusive and dispersive truncation errors occurring in many convective difference schemes, which will be discussed in Chapter 7.

3.3. Time-Differencing Schemes

The construction of finite-difference schemes for CFD requires both space and time differencing. Spatial differencing, especially in more than one space dimension, tends to produce lengthy and complicated algebraic expressions cluttered with subscripts and superscripts. When such expressions are written out in full detail, they can obscure the underlying relative simplicity of the time-advancement scheme. The structure and logic of the latter can be seen more clearly if the time differencing is explicitly displayed while the spatial differencing is temporarily suppressed. To this end, we shall denote the spatial difference approximation to an arbitrary quantity $Q(x)$ at the point $x = x_i$ by $\langle Q \rangle_i$, and similarly that at the points $x = x_{i\pm}$ by $\langle Q \rangle_{i\pm}$. The precise form of $\langle Q \rangle_i$ and/or $\langle Q \rangle_{i\pm}$ must of course eventually be specified to uniquely define the numerical scheme, but this can be deferred until after the time differencing has been defined. The time level at which the expression $\langle Q \rangle_i$ is evaluated will be denoted by a superscript in the usual way; e.g., $\langle Q \rangle_i^n$.

Equations (2.10), (2.12), and (2.13) respectively define forward, backward, and centered difference approximations to the time derivative

$\partial f/\partial t$. Each of those approximations gives rise to a corresponding basic time-differencing scheme for computing approximate solutions to the generic equation system of Eq. (1.1). As discussed in the Introduction, however, the individual terms in Eq. (1.1) will actually be considered separately, so we shall focus the discussion on Eq. (1.3) instead. As applied to Eq. (1.3), the basic forward, backward, and centered time-differencing schemes respectively take the form

$$\frac{\mathbf{v}_i^{n+1} - \mathbf{v}_i^n}{\Delta t} = \langle \mathcal{D}_\alpha(\mathbf{v}) \rangle_i^n \tag{3.4}$$

$$\frac{\mathbf{v}_i^{n+1} - \mathbf{v}_i^n}{\Delta t} = \langle \mathcal{D}_\alpha(\mathbf{v}) \rangle_i^{n+1} \tag{3.5}$$

$$\frac{\mathbf{v}_i^{n+1} - \mathbf{v}_i^n}{\Delta t} = \langle \mathcal{D}_\alpha(\mathbf{v}) \rangle_i^{n+} \tag{3.6}$$

in which the unspecified spatial differencing denoted by $\langle \cdots \rangle$ will generally differ from one time-differencing scheme to another. Most of the difference schemes considered in this book are special cases of Eqs. (3.4)–(3.6), which are respectively first-, first-, and second-order accurate in time by virtue of Eqs. (2.10), (2.12), and (2.13). These schemes are also referred to by various other names with which the reader should become familiar. Equation (3.4) is variously referred to as a *fully explicit, forward time*, or *Euler* scheme, while Eq. (3.5) is referred to as a *fully implicit, backward time*, or *backward Euler* scheme. Equation (3.6) is a *time-centered* scheme, and it too is implicit but not "fully" implicit, because the right member is evaluated halfway between the old and new time levels. The meaning of "explicit" and "implicit" in this context is that Eq. (3.4) can be trivially solved to obtain an *explicit* expression for the advanced or new-time value $\mathbf{v}_i^{n+1}$ in terms of the previous or old-time values $\mathbf{v}_i^n$, which have already been computed on the previous time step (or are given by the initial conditions if $n = 0$), and may therefore be regarded as known. In contrast, Eqs. (3.5) and (3.6) *implicitly* determine $\mathbf{v}_i^{n+1}$ in terms of $\mathbf{v}_i^n$, but they are not so easily solved for $\mathbf{v}_i^{n+1}$, which of course is necessary in order to advance the solution from one time level to the next. Indeed, solving for the advanced-time quantities in implicit schemes is nearly always tedious at best, and it often presents difficult challenges requiring sophisticated numerical analysis. Implicit schemes therefore tend to be significantly more complicated and troublesome than explicit schemes. The reason why implicit schemes are nevertheless widely used is that there are many situations in which their advantages outweigh their drawbacks, and in some situations (e.g., incompressible flow) there is no alternative because no suitable explicit schemes exist. Those advantages primarily pertain to *numerical stability* and associated considerations of computational efficiency, which will be discussed in the next chapter. An advantageous feature which is peculiar to the fully implicit scheme of Eq. (3.5) is that it provides the option of computing the steady-state solution in a single time step by setting Δt equal to a value so large that $(\mathbf{v}_i^{n+1} - \mathbf{v}_i^n)/\Delta t$ becomes negligible. When this is done, Eq. (3.5) reduces to $\langle \mathcal{D}_\alpha(\mathbf{v}) \rangle_i^{n+1} = 0$, which is just a spatial difference approximation to $\mathcal{D}_\alpha(\mathbf{v}) = 0$, with $\mathbf{v}$ replaced by $\mathbf{v}^{n+1}$. The latter equation is simply the steady-state limit to which Eq. (1.3) reduces when $\partial \mathbf{v}/\partial t = 0$. Of course, in order for this benefit to accrue to the full equation system of Eq. (1.1), all of the terms $\mathcal{D}_\alpha(\mathbf{v})$ therein would have to be differenced in a fully implicit manner; i.e., approximated at the advanced time level $n + 1$.

There is an infinite variety of implicit schemes, but most if not all of those employed in CFD share some general features that are consequences of the structure of the governing equations of fluid dynamics. Except in the case of source terms, the operator $\mathcal{D}_\alpha$ in Eqs. (1.3) and (3.4)–(3.6) is a spatial differential operator, and hence always contains either first or second spatial derivatives, and not infrequently both. The spatial difference approximations indicated by $\langle \cdots \rangle$ replace those derivatives by finite differences. As a result, the right members of Eqs. (3.4)–(3.6) generally involve the values of $\mathbf{v}$ not merely at the central mesh point i but at neighboring mesh points $j = i \pm 1$, and sometimes $i \pm 2$ or even $i \pm 3$ as well. Such coupling between neighboring mesh points is exemplified by Eqs. (2.3)–(2.5) and (2.9), or in the present context even better by Eq. (2.19).

The above observations have the following implications for implicit schemes, of which the fully implicit scheme of Eq. (3.5) serves as a convenient basis for the discussion. (Indeed, many if not most of the implicit schemes we shall consider can be regarded as approximations to fully implicit schemes, as will be discussed in the next section.) For any particular value of i, Eq. (3.5) involves not merely $\mathbf{v}_i^{n+1}$ but also the values of $\mathbf{v}_j^{n+1}$ at neighboring points j, and this prevents the immediate solution of Eq. (3.5) for $\mathbf{v}_i^{n+1}$ as can be done in the explicit Eq. (3.4). A simple example of such a scheme, which clearly illustrates the coupling to adjacent mesh points, is provided by Eq. (6.14) for the diffusion equation. The number of mesh points is normally quite large, so that Eq. (3.5) represents a large sparse system of coupled equations for the quantities $\mathbf{v}_i^{n+1}$ at all mesh points within the computational region. This equation system must somehow be solved on each time step in order to obtain the values of $\mathbf{v}_i^{n+1}$ and thereby advance the solution from one time level to the next. The meaning of "sparse" in this context is simply that the direct coupling between mesh points implied by Eq. (3.5) is limited to nearby neighbors, again as illustrated by Eq. (6.14), and this simplification can be exploited in solving for $\mathbf{v}_i^{n+1}$. Nevertheless, *all* the mesh points are indirectly coupled to each other, because each neighboring mesh point is itself coupled to slightly more distant neighboring mesh points, and so on until the chain of such connections extends throughout the entire mesh.

The problem of determining the quantities $\mathbf{v}_i^{n+1}$ is made even more difficult by the fact that the equations of fluid dynamics are nonlinear, so that $\mathcal{D}_\alpha(\mathbf{v})$ is in general a nonlinear operator and $\langle\mathcal{D}_\alpha(\mathbf{v})\rangle_i^{n+1}$ is consequently a nonlinear function of $\mathbf{v}_i^{n+1}$ and its nearby neighbors. In addition to being large and sparse, the coupled equation system that must be solved to obtain the values of $\mathbf{v}_i^{n+1}$ is therefore also nonlinear. There is no known general method for solving such equation systems directly. The only available methods of solution are iterative, of which the canonical paradigm is the venerable *Newton–Raphson method*. This method is an essential tool in many branches of numerical analysis, including CFD, and readers to whom it is new are hereby enjoined to become familiar with it. It converges in a single iteration when the equations are linear, and generally converges more rapidly when the nonlinearities are relatively weak. For this reason, its convergence can often be accelerated by means of nonlinear variable transformations that convert the original equations into equivalent equations that are more nearly linear (i.e., less strongly nonlinear). Many of the iterative methods employed in implicit CFD schemes are variants, refinements, or generalizations of the Newton–Raphson method. This is a complex, difficult, and highly specialized subject which lies outside the scope of this book and remains an active area of research.

The solution of nonlinear equation systems is much more difficult than that of linear equation systems of the same size, although the latter present significant challenges in their own right and also constitute a major branch of numerical analysis. Fortunately, various software packages are nowadays available for solving linear systems by both direct and iterative methods, such as LAPACK and ITPACK. (Those methods and packages are also useful in nonlinear Newton–Raphson-based solution schemes, which frequently require the exact or approximate solution of associated linear equation systems on each iteration.) These considerations provide an incentive to develop *linearly implicit* schemes in which the variables v_i^{n+1} only appear linearly, so that the large sparse equation system they satisfy becomes linear rather than nonlinear. When they are intelligently designed, such linearly implicit schemes provide many of the benefits of the corresponding fully implicit schemes, but they are computationally easier to deal with and solve. Linearly implicit schemes of this type can be constructed by suitably linearizing or otherwise approximating the nonlinear equations of the fully implicit scheme. There is no unique way to do this, and different approaches result in different linearly implicit schemes with different properties. Which approach is preferable depends on the particular problem or class of problems under consideration. Of course, in the special case when $\mathcal{D}_\alpha(\mathbf{v})$ is already linear in $\mathbf{v}$, the fully implicit scheme of Eq. (3.5) is already linearly implicit and requires no further linearization. This case is not uncommon, and occurs in particular for the pressure gradient terms and for viscous terms with constant viscosities.

Perhaps the most direct and straightforward way of constructing a linearly implicit approximation to Eq. (3.5) is to linearize $\langle \mathcal{D}_\alpha(\mathbf{v}) \rangle_i^{n+1}$ by replacing $\mathbf{v}^{n+1}$ therein with $\mathbf{v}^n + \Delta\mathbf{v}$, where $\Delta\mathbf{v} \equiv \mathbf{v}^{n+1} - \mathbf{v}^n$, expanding the result as a Taylor series in powers of $\Delta\mathbf{v}$, and truncating the expansion at the linear terms; i.e., discarding all terms of quadratic and higher order in $\Delta\mathbf{v}$. An alternative approach, which appears less systematic but in many cases turns out to be preferable, is to write out the quantity $\langle \mathcal{D}_\alpha(\mathbf{v}) \rangle_i^{n+1}$ in detail, carefully inspect it to identify all the places where $\mathbf{v}^{n+1}$ occurs, and then judiciously and selectively replace some of those occurrences by $\mathbf{v}^n$ until only linear occurrences of $\mathbf{v}^{n+1}$ remain. Such replacements are legitimate because $\mathbf{v}^n$ is a first-order accurate approximation to $\mathbf{v}^{n+1}$, provided of course that one does not mix the old and new time levels within difference approximations to spatial derivatives, as discussed in the previous chapter. Both of the above approaches are much easier to grasp and visualize within the context of specific examples, and they will be illustrated and discussed in more detail in Chapter 5.

Once suitable spatial and temporal difference approximations to the individual terms $\mathcal{D}_\alpha(\mathbf{v})$ have been selected, they must be combined to obtain a composite difference scheme for the full Eq. (1.1). It is only at this point that the composite scheme can actually be solved to obtain the advanced-time quantities $\mathbf{v}^{n+1}$. Such schemes commonly approximate some of the terms $\mathcal{D}_\alpha(\mathbf{v})$ implicitly and others explicitly. Composite schemes of this type may be referred to as *partially implicit*, and their implicitness can be either linear or nonlinear. Historically, the general practice has been to employ implict time differencing only where and to the degree that this is either necessary or highly advantageous. In a more nearly perfect universe, one would prefer not to use implicitness at all, since explicit schemes are much simpler and implicitness introduces significant complications as discussed above. However, since implicitness is sometimes unavoidable, it seems logical to keep it to a minimum. This was more or less the traditional philosophy by which many common partially implicit CFD schemes were developed.

3.4. Numerical Solution of Large Linear Systems

As discussed above, both fully and linearly implicit numerical schemes require the solution of large sparse systems of coupled linear equations. This subject area constitutes an important branch of numerical analysis to which numerous lengthy treatises have been devoted, so it lies well beyond the scope of this book. Fortunately, a detailed knowledge of this area is not a prequisite for all CFD work, but there are certain important aspects of it that all CFD practitioners should be aware of. We now proceed to briefly summarize some of those aspects, albeit with no pretense of completeness. In fact, the discussion consists of little more than a list of some important methods with which more advanced students will eventually be required to become familiar.

In contrast to nonlinear equation systems, linear equation systems are susceptible to direct methods of solution, of which the canonical paradigm is Gaussian elimination, which was actually discovered by the Chinese some two millenia before Gauss! However, direct methods are usually not computationally competitive with iterative methods for the large sparse linear systems that arise from linearly implicit CFD schemes. But this general assessment has one outstanding exception with which even beginning CFD students should become familiar, namely the Thomas algorithm for solving *tridiagonal* linear systems in one space dimension, in which the ith equation involves only the variables f_i and $f_{i\pm1}$. The Thomas algorithm is described in innumerable textbooks, so it would be wasteful of space and trees to discuss it here. It can be generalized to deal with equation systems of a related but slightly more complicated structure. It also finds use as a basic ingredient in certain iterative methods, as well as in alternating-direction implicit (ADI) schemes in two and three space dimensions. The latter schemes will be briefly discussed in Chapter 6.

As to iterative methods, it is also unfortunate, but not surprising, that the simplest and most easily understood iterative methods are not computationally competitive with more complicated and sophisticated methods that require a significantly greater investment of further study on the part of the user. But the simpler methods do work, at least most of the time, and there are many situations in which the benefits of simplicity outweigh the associated costs of relative inefficiency. The traditional simple iterative solution methods for linear systems of the present type are the Jacobi, Gauss–Seidel, and successive over-relaxation (SOR) methods, which are very easy to understand and program. They are generally considerably less efficient than conjugate gradient and conjugate residual methods, of which many competing variants exist. The latter methods are based on geometrical and linear algebra concepts which are actually relatively simple, but their details and structure are somewhat more complicated and intricate than those of the simpler methods. Mention must also be made of multigrid methods, which are designed to accelerate the otherwise relatively slow convergence rates of solution components with long wavelengths.

Chapter 4

NUMERICAL STABILITY

The concept of stability is fundamental in many branches of physics, and nowhere more so than in fluid dynamics, which abounds with physical instabilities. The essence of instability is the rapid growth of small perturbations or errors. Unfortunately, the time-marching numerical methods used in CFD are susceptible to a variety of unphysical instabilities that are purely numerical in origin. That is to say, they are artifacts of the finite-difference approximations and are not an intrinsic property of the differential equations. Such instabilities manifest themselves in the growth, from one time level to the next, of whatever inevitable small irregularities or errors are initially present in the solution, including roundoff errors. When it occurs, such growth is frequently very rapid and can result in the amplification of errors by orders of magnitude over just a few time steps. The accuracy of the numerical solution is thereby destroyed, and indeed the calculation usually terminates with computer overflow errors and cannot be continued. Such numerical instabilities must obviously be avoided if accurate CFD solutions are to be computed. This requires some means of analyzing finite-difference equations to determine their stability properties. Thus stability analysis takes its place beside truncation-error analysis as another and equally important test which proposed difference schemes must successfully pass before they are employed. It is also essential for CFD practitioners to familiarize themselves with the more common and well-known types of instabilities that have previously been identified and analyzed, and the situations in which they occur. Many of those common instabilities will be discussed in subsequent chapters.

There are three common methods for performing such stability analyses: the heuristic Fourier (or von Neumann) method, the matrix method, and the energy method. The Fourier and matrix methods provide necessary conditions for stability but are restricted to linear equations, so if the equations are not already linear then they must first be linearized before the analysis can be performed. The manner in which this is done is described in Sect. 4.2 below. The matrix method is the more rigorous of the two, particularly in allowing for the effects of boundary conditions, and is correspondingly more complicated and less convenient to use. In contrast, the energy method can be directly applied to some nonlinear equations but only provides sufficient conditions for stability, which may be unnecessarily restrictive. In practice, the Fourier method is by far the most useful and widely employed. It is very simple and easy to use, and serves as the standard reliable workhorse for performing the practical everyday CFD stability analyses that are routinely required. The usual practice of CFD would not be greatly hampered if the matrix and energy methods were unavailable, but it would be crippled without the Fourier method. We shall therefore not discuss the matrix and energy methods, but will restrict attention to the Fourier method.

4.1. Stability Conditions

Time-marching numerical methods can be separated into three basic categories: unconditionally unstable, conditionally stable, and unconditionally stable. The "condition" to which these terms refer is a constraint on the time step Δt. Unconditionally unstable schemes are unstable for all values of Δt, and hence are unusable, at least in isolation, and must or at least should be rejected. They can, however, be insidious, because in composite numerical schemes their instability is sometimes masked or suppressed by numerical damping introduced by other parts of the scheme. Some composite schemes of this type are actually legitimate and useful, but as a general rule schemes that rely on cancellation of errors are undesirable and should be avoided. The reason is obvious: the relative magnitudes of the various errors are usually problem-dependent, so that situations will eventually be encountered in which the desired cancellation is incomplete or does not occur at all. It is therefore a wise general policy to require each part of the scheme to be both consistent and stable in isolation, independently of the other parts.

Conditionally stable schemes are typically stable for sufficiently small values of Δt, and unstable otherwise. Such schemes are therefore useful only when and if an associated *stability condition* or restriction of the form $\Delta t \leq \Delta t^*$ is satisfied, where Δt^* is the largest value of Δt for which the scheme is stable. We shall encounter numerous examples of such stability conditions in succeeding chapters. The stability limit $\Delta t^{\star}$ depends on the particular scheme and is not simply a constant, but ordinarily depends on Δx as well as the values of the various parameters (e.g., viscosity, sound speed, etc.) and dependent variables (e.g., density, velocity, etc.) in the problem. Except in pathological cases, $\Delta t^{\star}$ tends to zero in the limit as $\Delta x \rightarrow 0$. Almost all stable explicit schemes are conditionally stable in this sense. It frequently happens that their associated stability conditions require the use of much smaller values of Δt than would otherwise be required to resolve the physical phenomena of interest. Situations of this type are the main motivation for the use of implicit schemes, which can often be constructed in such a way that they have much less restrictive stability conditions. Indeed, many implicit schemes are actually unconditionally stable; i.e., stable for all finite values of Δt. Unconditional stability is a highly desirable property, but it rarely provides the freedom to use an arbitrarily large time step because Δt is also subject to accuracy restrictions. Of course, there is no sharp boundary between accuracy and inaccuracy as there is between stability and instability, but the numerical solution cannot be expected to be physically accurate unless Δt is small compared to the physical characteristic times in the original differential equations. In most cases, explicit stability limits turn out to be of the same order of magnitude as those physical characteristic times, so that the accuracy limits on implicit schemes are often comparable to the stability limits on explicit schemes. Why then are implicit schemes nevertheless of interest? For one thing, they provide the option to continue the calculation expeditiously by accepting a temporary reduction in its accuracy, especially in localized regions of space and/or time. But more importantly, they provide a way of sacrificing the accuracy of particular solution features that may not be of interest but would otherwise require the use of unacceptably small values of Δt (e.g., sound waves at very low Mach number), or quantities that are so small as to be insignificant or negligible. For example, if the mass fraction y of a trace species in a gas mixture is considered negligible (i.e., it might as well be zero) as long as $y < 10^{-5}$, then if its true value is $y = 10^{-8}$ it can be in error by a factor of 1,000 without adverse effect. Implicit schemes are especially useful in removing excessively restrictive explicit stability conditions due to *stiffness*, which typically gives rise to uninteresting, irrelevant, and/or insignificant solution features with very small amplitudes but very short time scales. We shall return to this point in Sects. 5.1.4 and 8.1.2.

Strictly speaking, stability is not a property of the equations alone, but rather of particular solutions to those equations. For example, the Navier–Stokes equations describe an extremely wide variety of fluid flows, some of which are physically stable while others are not. However, this distinction is less important for difference schemes, which are intended to be of general applicability and hence should ideally be required to be numerically stable for all solutions, in the loose sense that no numerical solution should ever be more unstable than the physical solution it approximates. This requirement raises the nontrivial question of how to distinguish between physical and numerical instabilities in situations where the former occur, or are suspected of occurring. Rigorous mathematical criteria for numerical stability have been developed for this purpose, but in everyday practice they are seldom required. Routine stability analyses of finite-difference schemes normally restrict attention to situations in which the corresponding differential equations have simple stable solutions, so that any instability that occurs can be unambiguously attributed to the difference scheme. Even with this restriction, it is not feasible to examine the stability of all possible numerical solutions of this type. In practice, stability analyses are usually performed only for the simplest unperturbed solutions, namely uniform steady-state solutions whose space and time derivatives and differences vanish in both the differential and difference equations. (Such solutions are normally physically stable, but not always, so if there is any doubt about this it should be confirmed by a differential stability analysis.) Difference equations that are stable for such solutions are then provisionally presumed to be stable for general use in the absence of other evidence to the contrary.

4.2. Linearization

Most stability analyses of either differential or difference equations are based on *linearization*. This process proceeds in two steps. First, $\mathbf{v}$ is replaced everywhere in the equations by $\mathbf{v}_0 + \delta\mathbf{v}$, where $\mathbf{v}_0$ is the unperturbed solution whose stability is being analyzed, and $\delta\mathbf{v}$ is the perturbation, which is initially supposed to be very small. Second, the resulting equations are systematically linearized by expanding them in powers of $\delta\mathbf{v}$ and discarding all terms proportional to powers of $\delta\mathbf{v}$ higher than the first. This step is of course unnecessary when the original equations are already linear in $\mathbf{v}$. Otherwise, linearization represents an approximation that is valid as long as $\delta\mathbf{v}$ is sufficiently small, which can always be ensured simply by taking its initial value to be as small as necessary. Even if $\delta\mathbf{v}$ subsequently increases due to an instability, which indeed is precisely what the analysis seeks to determine, it will remain sufficiently small for at least some brief interval of time, during which its rate of growth or decay can be ascertained.

By construction, the above procedure results in a linear system of equations for $\delta \mathbf{v}$. Those equations determine $\delta \mathbf{v}$ as a function of either (x, t) in the differential case, or the corresponding indices (i, n) in the present finite-difference context. In general, the coefficients in those equations depend on the unperturbed solution $\mathbf{v}_0$, and hence are constants when the latter is a uniform steady-state solution of the original equations. Otherwise those coefficients are variable and depend on (i, n) via their dependence on $\mathbf{v}_0$. The terms in the linearized equations that do not involve $\delta \mathbf{v}$ drop out by virtue of the fact that $\mathbf{v}_0$ satisfies the original equations. If the latter were already linear in $\mathbf{v}$, then it is easy to verify that $\delta \mathbf{v}$ satisfies those same equations *sans* any inhomogeneous terms independent of $\mathbf{v}$, which provides a convenient shortcut to the equations determining $\delta \mathbf{v}$.

In actual practice, the above procedure is quite simple and straightforward, as we shall illustrate by applying it to the forced inviscid Burgers equation in one space dimension:

$$\frac{\partial v}{\partial t} + v \frac{\partial v}{\partial x} = S \tag{4.1}$$

where $S = S(x, t)$ is a specified inhomogenous forcing function which is independent of the dependent variable $v(x, t)$. As instructed above, we first replace v by $v_0 + \delta v$ to obtain

$$\frac{\partial v_0}{\partial t} + \frac{\partial (\delta v)}{\partial t} + (v_0 + \delta v) \left(\frac{\partial v_0}{\partial x} + \frac{\partial (\delta v)}{\partial x} \right) = S \tag{4.2}$$

But the unperturbed solution $v_0(x, t)$ itself satisfies Eq. (4.1), so that

$$\frac{\partial v_0}{\partial t} + v_0 \frac{\partial v_0}{\partial x} = S \tag{4.3}$$

Combining Eqs. (4.2) and (4.3) and neglecting the nonlinear term of order δv^2, we obtain

$$\frac{\partial (\delta v)}{\partial t} + v_0 \frac{\partial (\delta v)}{\partial x} + \delta v \frac{\partial v_0}{\partial x} = 0 \tag{4.4}$$

Note that the inhomogeneous term $S(x, t)$ has dropped out as promised, so that Eq. (4.4) is a linear homogeneous equation that determines the perturbation $\delta v(x, t)$. In the common special case in which v_0 is a uniform steady-state solution, $\partial v_0 / \partial x = 0$ and Eq. (4.4) further reduces to

$$\frac{\partial(\delta v)}{\partial t} + v_0 \frac{\partial(\delta v)}{\partial x} = 0 \qquad (4.5)$$

which is now a homogeneous linear equation with constant coefficients for δv, and is easily solved in terms of Fourier modes as discussed below.

Once the linearization has been performed, the condition for stability is that there should be no amplification of the perturbation from one time step to the next; i.e.,

$$|\delta \mathbf{v}^{n+1}| \leq |\delta \mathbf{v}^n| \qquad (4.6)$$

for all permissible choices of the initial perturbation $\delta \mathbf{v}^0$, where $|\delta \mathbf{v}|$ is some appropriate measure of the amplitude, magnitude, modulus, norm, or absolute value of $\delta \mathbf{v}$. We reemphasize that this definition of numerical stability is based on the assumption that there is no physical instability present which might result in a legitimate physical growth of $\delta \mathbf{v}$. Of course, Eq. (4.6) cannot be evaluated or imposed until the linearized difference equations have first been solved to obtain $\delta \mathbf{v}^n$, and this is where the Fourier (or matrix) method comes in. The great virtue of the Fourier method is that it provides a simple and convenient way of representing and varying $\delta \mathbf{v}^0$ and of computing the resulting $\delta \mathbf{v}^n$, thereby enabling the stability restriction on Δt to be inferred from Eq. (4.6).

4.3. Heuristic Fourier Stability Analysis

The Fourier (or von Neumann) method is a heuristic method which is lacking in rigor, at least as it is usually employed, but a vast wealth of empirical evidence accumulated in countless calculations over many decades of experience provides ample evidence that it generally provides essentially correct stability conditions for a wide variety of numerical schemes. The method has therefore proved to be very reliable for most practical purposes. The essence of the method is to consider perturbations of the form of a single Fourier mode; i.e.,

$$\delta \mathbf{v}_i^n = \hat{\mathbf{v}} \exp\{\iota(\kappa x - \omega t)\} = \hat{\mathbf{v}} \exp\{\iota(\kappa i \Delta x - \omega n \Delta t)\} \qquad (4.7)$$

where $\iota = \sqrt{-1}$, $\kappa = 2\pi/\lambda$ is the wavenumber of the perturbation, λ is its wavelength, and ω is its angular frequency. It is convenient to define the complex *amplification factor* $\xi \equiv \exp\{-\iota\omega\Delta t\}$, so that Eq. (4.7) becomes

$$\delta\mathbf{v}_i^n = \hat{\mathbf{v}}\,\xi^n \exp\{\iota\kappa i\Delta x\} \tag{4.8}$$

It can be shown that an arbitrary perturbation can be expressed as a linear superposition of such modes. Why then did we not write $\delta\mathbf{v}_i^n$ as a summation over such modes? The reason is the essential simplifying feature of the analysis: it can further be shown that when the linear equations for $\delta\mathbf{v}_i^n$ have constant coefficients and periodic boundary conditions, or boundaries at infinity, the individual Fourier modes evolve independently of each other, so that each individual mode is itself a solution of the equations and can consequently be considered separately. Thus the stability of each mode can be examined in isolation from the others. Each mode κ thereby gives rise to its own stability condition via Eq. (4.6), which now simply reduces to $|\xi(\kappa)| \leq 1$, where $\xi(\kappa)$ is the amplification factor for that particular value of κ. The stability condition for the scheme as a whole is that it must be stable for all initial perturbations; i.e., all Fourier modes. The overall stability condition then takes the form

$$|\xi(\kappa)| \leq 1 \quad \text{for all } \kappa \tag{4.9}$$

which requires

$$\max_{\kappa} |\xi(\kappa)| \leq 1 \tag{4.10}$$

In actual practice, it is frequently more convenient to impose Eq. (4.9) rather than Eq. (4.10). In either case, the wavenumber κ varies over all of its physically relevant values. The shortest wavelength that can be supported by the mesh is $\lambda = 2\Delta x$, so when the boundaries are at infinity the relevant range of κ is $0 < \kappa < \pi/\Delta x$. Either Eq. (4.9) or (4.10) normally reduces to a restriction on Δt, since $\xi(\kappa)$ is also a function of Δt, Δx, other parameters in the equations, and in general the values of the dependent variables in the unperturbed solution.

According to Eqs. (4.9) and (4.10), all allowed wavelengths must be examined to establish stability. Conversely, however, a single unstable wavelength suffices to establish instability. The least stable or most unstable wavelength is often the shortest; i.e., $\kappa = \pi/\Delta x$. In situations where the algebra required to examine the behavior of all wavelengths is intractable or unduly tedious, it is frequently useful to derive *necessary* stability conditions by restricting attention to $\kappa = \pi/\Delta x$, and perhaps a few other values of κ for which $\xi(\kappa)$ can readily be evaluated. In many cases, the resulting necessary stability conditions turn out to be sufficient as well, although one has no assurance that this will occur in any particular instance.

All that remains is to determine ξ as a function of κ, which is done by substituting Eq. (4.8) into the linearized difference scheme and solving for ξ. This whole process is actually very straightforward for simple difference schemes, and it will be repeatedly illustrated in subsequent chapters.

The Fourier method as described above, with its attendant restrictions to constant coefficients and periodic boundary conditions, can actually be made quite rigorous. In practice, however, the same type of analysis is often heuristically employed in a localized manner even when the unperturbed solution is not uniform in space and/or steady in time (and for other types of boundary conditions), so that the above restrictions are not satisfied. This is done by ignoring the boundary conditions and regarding the computational region as infinite, and "freezing" the variable coefficients in the linear equations for $\delta\mathbf{v}_i^n$ by replacing them with the constant values obtained by evaluating the unperturbed solution $\mathbf{v}_0$ at some particular mesh point $i = i_0$ and time level $n = n_0$. One thereby obtains a system of equations with constant coefficients, to which the Fourier method as described above may then be applied. The rationale for this mathematically unjustified procedure is based on two complementary considerations: (a) the unperturbed solution $\mathbf{v}_0$ is normally a slowly varying function of (x, t) or (i, n), and hence is essentially constant in the immediate neighborhood of $(i, n) = (i_0, n_0)$, and (b) most numerical instabilities are highly localized phenomena in which wavelengths of order $2\Delta x$ (i.e., values of $\kappa \sim \pi/\Delta x$) grow the fastest and hence determine the overall stability condition of Eq. (4.9) or (4.10). Of course, the latter condition will now depend, in general, on (i_0, n_0), so the last step in the heuristic procedure is to allow i_0 to vary over all values of i in the computing mesh, so that the frozen coefficients similarly vary over all their values occurring in the mesh at time level n_0. The most restrictive stability condition thereby obtained then becomes the final overall stability condition for n close to n_0. For example, the common convective stability condition $|u|\Delta t/\Delta x \leq 1$, which will be derived and discussed in Chapter 7, is most restrictive when the fluid speed $|u|$ is largest, and hence should be evaluated using the maximum value of $|u|$ occurring in the mesh.

In common CFD practice, heuristic Fourier stability analyses of the type described above constitute the majority of applications of the Fourier method. Since such analyses are far from rigorous, the resulting stability restrictions on Δt cannot be expected to be precise in practical calculations. As a general rule, the empirical stability limits on Δt that are actually observed in calculations tend to be smaller than those resulting from the heuristic Fourier analysis by correction factors of order unity. It is therefore prudent and customary to manually insert such correction or safety factors into the resulting stability conditions. Such safety factors are typically 0.5 or less, and values of 0.25 are not uncommon, but values on the order of 0.1 or less are rarely required and suggest that something may have been overlooked in the analysis.

Fourier stability analyses for the composite difference schemes used to obtain numerical solutions to the full Eq. (1.1) in its entirety are generally too complicated to be carried out analytically. When such is the case, the stability condition for the composite scheme as a whole can usually be reasonably well approximated in terms of the stability conditions for the constituent terms and schemes of which it is comprised, such as Eqs. (3.4)–(3.6). If the stability conditions for those individual constituents are expressed in the form $\Delta t \leq \Delta t_\alpha^\star$ (where $\Delta t_\alpha^\star$ may include a safety factor of order unity as described above), then the condition

$$\sum_\alpha \frac{\Delta t}{\Delta t_\alpha^\star} \leq 1 \qquad (4.11)$$

frequently provides a very useful approximation to the stability condition for the composite scheme as a whole. Equation (4.11) may be rewritten as

$$\Delta t \leq \left(\sum_\alpha \frac{1}{\Delta t_\alpha^\star} \right)^{-1} \qquad (4.12)$$

in which another safety factor of order unity can be inserted if desired. Note that this condition is more restrictive than the obvious intuitive condition $\Delta t \leq \min_\alpha \Delta t_\alpha^\star$, and it is generally more accurate as well. A vast wealth of accumulated computational experience provides ample empirical evidence that the accuracy of Eq. (4.12) is usually quite sufficient for practical purposes. Indeed, in cases where the composite scheme is sufficiently simple that its full stability analysis can be carried out analytically, Eq. (4.12) is often found to be exact.

4.4. Nonlinear Instabilities and Chaos

The heuristic Fourier method accounts for the vast majority of observed stability behavior in finite-difference schemes, at least to within correction factors of order unity as discussed above. Occasionally, however, anomalous behavior is observed, such as the relatively slow and apparently unstable growth of unphysical solution features in situations where the linear Fourier analysis predicts stability. Such behavior has often been vaguely attributed to intrinsically "nonlinear instabilities," the existence and nature of which have been controversial. It has sometimes been suggested that, at least in some cases, such behavior may well be related to the ubiquitous phenomenon of chaos. This connection is suggested by the fact that by their very nature, time-marching finite-difference schemes determine $\mathbf{v}^{n+1}$ from $\mathbf{v}^n$, and can therefore be regarded as high-dimensional iterated nonlinear mappings of the general form $\mathbf{v}^{n+1} = \mathcal{F}(\mathbf{v}^n)$. Low-dimensional iterated maps have been extensively studied in the chaos literature, and numerous examples are known in which they manifest chaos and strange attractors. It would not be surprising in the least if their higher-dimensional finite-difference cousins were to sometimes exhibit similar chaotic behavior. Indeed, it would be somewhat surprising if they did not, and it is relatively easy to imagine this type of behavior manifesting itself in mysterious symptoms of the type sometimes attributed to "nonlinear instabilities." If so, this inherently nonlinear behavior would be difficult or impossible to diagnose by more traditional methods of analysis. These questions have been pursued by several investigators, but fall outside the scope of this book and the expertise of its author.

Chapter 5

SOURCE TERMS

The simplest type of terms occurring in $\partial \mathbf{v}/\partial t$ are terms that involve the values of the dependent variables $\mathbf{v}$ at the same space-time point $(\mathbf{x}, t)$, but not their spatial derivatives. We shall refer to such terms as *source terms*, even though that terminology is also often applied to terms that are known or prescribed functions of the *independent* variables $(\mathbf{x}, t)$. Source terms of the latter type are excluded from the present discussion, since they are essentially trivial and are easily treated in a straightforward manner. The dependence of the source terms of present interest on the dependent variables can be either linear or nonlinear. In fluid dynamics, the most common physical origins of such terms are frictional forces, heat transfer by Newtonian heating/cooling or radiative transfer, and chemical reactions. Since such terms do not involve spatial derivatives, their finite-difference approximations likewise do not involve spatial differences and the associated direct coupling between neighboring mesh points of the type exemplified by Eq. (2.9). This lack of coupling makes source terms much easier to deal with, and allows us to suppress the spatial index i in discussing them. As we shall see, however, their numerical treatment also provides valuable insight into explicit vs. implicit time differencing for certain terms involving spatial derivatives as well, particularly diffusion terms.

5.1. Linear Source Terms

We begin by considering linear source terms, which are mathematically trivial and yet provide a great deal of insight into the proper numerical treatment of source terms in general. The simplest example of a linear source term is provided by the differential equation

$$\frac{\partial f}{\partial t} = \beta f \tag{5.1}$$

the analytical solution of which is simply

$$f(t) = f(0) \exp\{\beta t\} \tag{5.2}$$

Thus $|\beta|$ has the significance of an inverse time constant or characteristic time. The behavior of this solution is qualitatively different depending on whether β is positive or negative. For $\beta > 0$, $f(t)$ rapidly diverges without bound as $t \to \infty$, while for $\beta < 0$ $f(t)$ remains bounded and in fact exponentially decays to zero. Moreover, the time dependence is monotonic in both cases.

5.1.1. *Fully explicit time differencing*

We first consider the consequences of approximating Eq. (5.1) using the fully explicit time-differencing scheme of Eq. (3.4):

$$\frac{f^{n+1} - f^n}{\Delta t} = \beta f^n \tag{5.3}$$

Owing to the absence of spatial derivatives, this equation can easily be solved exactly by rewriting it as

$$f^{n+1} = (1 + \beta \Delta t) f^n \tag{5.4}$$

the solution of which is obviously

$$f^n = (1 + \beta \Delta t)^n f^0 \tag{5.5}$$

This equation, and similar equations following, commits an unfortunate notational inconsistency, since the superscript n denotes a time level in the left member and an exponent in the right member. This should cause no confusion, however, because $1 + \beta \Delta t$ does not involve f, so it would obviously not be sensible to interpret its superscript as a time level.

Exercise 5.1. Show that in the limit as $\Delta t \to 0$ and $n \to \infty$ at fixed $t = n\Delta t$, Eq. (5.5) properly converges to the exact analytical solution Eq. (5.2).

We now proceed to examine the qualitative behavior of Eq. (5.5) for large n, in particular how well it represents the qualitative behavior of

Eq. (5.2) for large t. To this end, we rewrite Eq. (5.2) in the form

$$f(n\Delta t) = \exp\{\beta n\Delta t\}f(0) = (\exp\{\beta\Delta t\})^n f(0) \qquad (5.6)$$

Comparison with Eq. (5.5) then shows that the difference scheme effectively approximates the factor $\exp\{\beta\Delta t\}$ by $1 + \beta\Delta t$, which obviously is quantitatively accurate only when $\beta\Delta t \ll 1$.

Consider first the case $\beta > 0$. In that case Eq. (5.5) predicts that f^n diverges monotonically without bound as $n \to \infty$, in qualitative agreement with Eq. (5.2) or (5.6). However, it does so at a slower rate, since $1 + \beta\Delta t < \exp\{\beta\Delta t\}$ for positive β. The fully explicit scheme of Eq. (5.3) is therefore "stable" in the sense that its solution is *less unstable* than the exact analytical solution of the original differential equation for any Δt. Thus when $\beta > 0$, the fully explicit scheme is accurate for small Δt, exhibits the same qualitative behavior as the exact analytial solution for large Δt, and is unconditionally stable relative to the analytical solution.

Now consider the case $\beta < 0$, in which the exact analytical solution Eq. (5.2) monotonically decays to zero as $t \to \infty$ and exhibits no physical instability. We may therefore analyze the numerical stability of the difference scheme by means of the heuristic Fourier method described in the previous chapter. Since Eq. (5.3) is already linear and contains no inhomogeneous terms, the perturbation $\delta f_i^n = \hat{f}\xi^n \exp\{\iota\kappa i\Delta x\}$ satisfies precisely the same equation. Common factors of $\hat{f}\xi^n \exp\{\iota\kappa i\Delta x\}$ cancel out, and we thereby obtain $(\xi - 1)/\Delta t = \beta$, so that

$$\xi = 1 + \beta\Delta t = 1 - |\beta|\Delta t \qquad (5.7)$$

since $\beta < 0$. The stability condition $|\xi| \leq 1$ then reduces to $1 - |\beta|\Delta t \geq -1$, or

$$\Delta t \leq 2/|\beta| \qquad (5.8)$$

When this condition is violated, the numerical solution Eq. (5.5) diverges for large n, in qualitative disagreement with the analytical solution Eq. (5.2). Conversely, when $\Delta t < 2/|\beta|$ the numerical solution properly decays to zero for large n, but this in itself is not sufficient to ensure qualitatively correct behavior. Inspection of Eq. (5.5) shows that when $\Delta t > 1/|\beta|$, the numerical solution exhibits unphysical oscillations: it overshoots zero and takes on unphysical negative values on alternate time steps, in qualitative disagreement with the monotonic decay of the analytical solution. Thus we see that the more restrictive condition $\Delta t \leq 1/|\beta|$ must be satisfied in order for the numerical solution of the fully explicit numerical scheme to exhibit the same qualitative decay behavior as the analytical solution when $\beta < 0$.

5.1.2. *Fully implicit time differencing*

So much for the fully explicit scheme. We now proceed to approximate Eq. (5.1) by the fully implicit scheme of Eq. (3.5), which results in the difference equation

$$\frac{f^{n+1} - f^n}{\Delta t} = \beta f^{n+1} \tag{5.9}$$

or equivalently

$$f^{n+1} = \frac{f^n}{1 - \beta \Delta t} \tag{5.10}$$

the solution of which is

$$f^n = \frac{f_0}{(1 - \beta \Delta t)^n} \tag{5.11}$$

Comparison with Eq. (5.6) shows that the fully implicit scheme effectively approximates $\exp\{\beta \Delta t\}$ by $(1 - \beta \Delta t)^{-1}$. It is easy to verify that in the limit as $\Delta t \to 0$ at fixed $t = n\Delta t$, Eq. (5.11) also properly converges to the exact analytical solution Eq. (5.2) or (5.6).

As before, we now proceed to examine the qualitative behavior of Eq. (5.11) for large n for both $\beta > 0$ and $\beta < 0$. Consider first the case $\beta > 0$, in which the analytical solution diverges monotonically and exponentially as $t \to \infty$. For $\beta \Delta t \leq 1$, Eq. (5.11) exhibits similar behavior, except that it now diverges *faster* than the analytical solution, since in this case it can easily be shown that $(1 - \beta \Delta t)^{-1} > \exp\{\beta \Delta t\}$. Indeed, the numerical solution diverges infinitely fast when $\beta \Delta t = 1$. The fully implicit scheme is therefore numerically "unstable" for $\beta \Delta t \leq 1$ in the sense that it is more unstable (i.e., grows faster) than the analytical solution. When $\beta \Delta t > 1$ the behavior of the numerical solution becomes even worse: it exhibits unphysical oscillatory behavior and changes sign on alternate time steps, in qualitative disagreement with the monotonically increasing exact solution. Thus, although the fully implicit scheme is consistent and consequently accurate for sufficiently small Δt, it is nevertheless unstable relative to the analytical solution when $\beta \Delta t \leq 1$, and it produces completely unacceptable qualitative behavior when $\beta \Delta t > 1$. We therefore conclude that when $\beta > 0$, the fully implicit scheme manifests unacceptable behavior for all values of Δt and hence should not be used. This conclusion contradicts the common misconception in some quarters of the CFD community that fully implicit schemes are universally ideal and should always be used whenever their associated complications (see Chapter 3) are tolerable. The above analysis shows that such a view is deeply fallacious, and emphasizes the desirability of examining the behavior of all the obvious simple special cases before embracing general recommendations.

We now consider the case $\beta < 0$, in which the analytical solution of Eq. (5.2) decays monotically to zero as $t \to \infty$ and exhibits no physical instability. A stability analysis entirely similar to that performed for the fully explicit scheme then yields

$$\xi = \frac{1}{1 - \beta \Delta t} = \frac{1}{1 + |\beta| \Delta t} \qquad (5.12)$$

This obviously satisfies the stability condition $\xi \leq 1$ for any value of Δt, so the fully implicit scheme is unconditionally stable when $\beta < 0$. Moreover, inspection of Eq. (5.11) shows that in the present case the numerical solution monotonically decays to zero as $n \to \infty$, in qualitative agreement with the exact analytical solution, although the latter goes to zero more rapidly. We therefore conclude that in all obvious respects, the fully implicit scheme is very well suited to the case $\beta < 0$, in contrast to its unacceptable behavior for $\beta > 0$.

The above observations suggest that fully implicit schemes are likely to be well suited to the description of stable physical processes, especially those involving irreversible decay, and conversely ill suited to processes involving the growth of physical instabilities. We shall see in the next chapter that this expectation is confirmed in the case of diffusion, which describes the irreversible decay of spatial inhomogeneities.

5.1.3. *Centered time differencing*

The fully explicit and fully implicit schemes considered above are both first-order accurate in time. We now proceed to examine the consequences of approximating Eq. (5.1) by the time-centered difference scheme of Eq. (3.6), which is second-order accurate. This results in the difference equation

$$\frac{f^{n+1} - f^n}{\Delta t} = \beta f^{n+} \qquad (5.13)$$

However, f^{n+} is not a primary computed variable like f^n, so we must further approximate it in terms of the latter. We shall adopt the obvious simple approximation

$$f^{n+} \approx \tfrac{1}{2}(f^n + f^{n+1}) \tag{5.14}$$

so that Eq. (5.13) is replaced by

$$\frac{f^{n+1} - f^n}{\Delta t} = \tfrac{1}{2}\beta(f^n + f^{n+1}) \tag{5.15}$$

This is the well-known *Crank–Nicholson scheme*, although that terminology is more commonly associated with its application to the diffusion equation, which will be discussed in the next chapter.

Exercise 5.2. Show that Eqs. (5.14) and (5.15) are both second-order accurate by means of truncation-error analyses.

Equation (5.15) can be rewritten as

$$f^{n+1} = \left(\frac{1 + \beta\Delta t/2}{1 - \beta\Delta t/2}\right) f^n \tag{5.16}$$

the solution of which is

$$f^n = \left(\frac{1 + \beta\Delta t/2}{1 - \beta\Delta t/2}\right)^n f^0 \tag{5.17}$$

Exercise 5.3. Verify that the numerical solution Eq. (5.17) converges to the analytical solution Eq. (5.2) or (5.6) in the limit as $\Delta t \to 0$ at fixed $t = n\Delta t$.

We first examine the behavior of Eq. (5.17) for $\beta > 0$, in which the analytical solution diverges monotonically for large t or n. When $\beta\Delta t \leq 2$, Eq. (5.17) exhibits similar behavior, but again can be shown to diverge faster than the analytical solution Eq. (5.6), at a rate that becomes infinite when $\beta\Delta t = 2$. The Crank–Nicholson scheme is therefore again numerically "unstable" for $\beta\Delta t \leq 2$ in the sense that it is more unstable (i.e., grows faster) than the analytical solution, albeit to a lesser degree than the fully implicit scheme. When $\beta\Delta t > 2$, Eq. (5.17) exhibits the entirely unphysical oscillatory behavior of changing sign on alternate time steps, in qualitative disagreement with the monotonically increasing exact solution. The behavior of the Crank–Nicholson scheme therefore exactly parallels that of the fully implicit scheme, the main difference being that the boundary between the two types of unacceptable behavior has been shifted from $\beta\Delta t = 1$ to $\beta\Delta t = 2$. We therefore conclude that when $\beta > 0$, the Crank–Nicholson scheme, like the fully implicit scheme, manifests unacceptable behavior for all values of Δt and hence should not be used.

Finally, we consider the behavior of Eq. (5.17) for $\beta < 0$, in which the analytical solution is stable and decays monotonically to zero. In this case, another routine stability analysis of Eq. (5.15) yields

$$\xi = \left(\frac{1 + \beta\Delta t/2}{1 - \beta\Delta t/2}\right) = \left(\frac{1 - |\beta|\Delta t/2}{1 + |\beta|\Delta t/2}\right) \tag{5.18}$$

Since ξ is real, the stability condition $|\xi| \leq 1$ reduces to $-1 \leq \xi \leq 1$, and both of those inequalities are easily seen to be satisfied. The Crank–Nicholson scheme is therefore unconditionally stable for $\beta < 0$. In contrast to the fully implicit scheme, however, it does not exhibit qualitatively correct behavior for all values of Δt. Inspection of Eq. (5.16) shows that when $\beta\Delta t > 2$, the numerical solution exhibits unphysical oscillations by overshooting zero to negative values on alternate time steps, in qualitative disagreement with the monotonic decay of the analytical solution. This is the same type of unphysical behavior exhibited by the fully explicit scheme when $\beta\Delta t > 1$, so the Crank–Nicholson scheme delays the onset of this behavior by doubling the value of Δt at which it first occurs. However, its qualitatively unphysical solution behavior for $\beta\Delta t > 2$ effectively negates the benefits of its unconditional stability.

In summary, the time-centered Crank–Nicholson scheme, like the fully implicit scheme, is actually inferior to the fully explicit scheme when $\beta > 0$, and indeed is essentially unsuitable for use in that case. Moreover, in spite of its second-order accuracy, its qualitative solution behavior for $\beta < 0$ is decidedly inferior to that of the fully implicit scheme. Indeed, it does not even enjoy a significant advantage over the fully explicit scheme in that case, since the restriction on Δt required to prevent unphysical oscillatory behavior is the same as the explicit stability limit of Eq. (5.8). We therefore conclude that the Crank–Nicholson scheme is of little interest as a numerical method for treating source terms of either sign, in spite of its second-order accuracy. In contrast, the first-order accurate fully explicit and fully implicit schemes are very well suited to source terms with $\beta > 0$ and $\beta < 0$, respectively, for which they provide eminently satisfactory treatments. (When the sign of β is not known *a priori,* one can simply monitor it and switch between the fully explicit and fully implicit schemes whenever it changes.) This situation serves as a concrete illustration of the remarks made in Chapter 2 to the effect that truncation-error analysis and order of accuracy are only relevant when Δt and/or Δx are sufficiently small that the Taylor series converges rapidly. This is not always the case in practical engineering calculations, in which it is desirable for the numerical solution to remain qualitatively reasonable even when the space or time discretization is not fine enough to accurately resolve the smallest space or time scales present in the exact solution. However, in such situations it is obviously important to have plausible grounds for believing that the resulting inaccuracies in inadequately or poorly resolved solution features are irrelevant, and conversely that the features of interest are likely to be sufficiently accurate for the purpose at hand.

5.1.4. *Further remarks on explicit vs. implicit schemes*

The preceding analysis would seem to suggest that *ceteris paribus*, implicit schemes are better suited to the numerical approximation of processes involving decay or relaxation, while conversely explicit schemes are better suited to processes involving growth or physical instability. But if this is indeed a valid general tendency, what is its basis? This tendency can readily be understood by observing that compared to centered time differencing, fully explicit schemes approximate time derivatives in terms of the prevailing conditions $\frac{1}{2}\Delta t$ earlier, while fully implicit schemes approximate time derivatives in terms of the conditions that they anticipate will prevail $\frac{1}{2}\Delta t$ later. Fully explicit schemes are consequently likely to overestimate time derivatives and decay rates in problems involving decay, thereby presenting a danger that the solution will unphysically overshoot its proper steady-state or equilibrium value. This must be prevented by imposing appropriate restrictions on Δt, which are sometimes onerous. Conversely, explicit schemes are likely to underestimate time derivatives and growth rates in problems involving growth, thereby numerically tempering the growth rate. This of course introduces some inaccuracy, but is generally preferable to the opposite inaccuracy of artificially exaggerating the growth rate.

Fully implicit schemes tend to produce the opposite behavior. Since they approximate time derivatives in terms of the conditions that they anticipate will exist a short time later, they are likely to underestimate time derivatives and decay rates in problems involving decay. This again introduces some inaccuracy, but prevents overshoots and ensures that the numerical solution comes in for a soft landing, so to speak. Conversely, by looking forward in time fully implicit schemes are likely to overestimate time derivatives and growth rates in problems involving growth, thereby making the numerical solution more unstable than the analytical one, just as we found in Sect. 5.1.2.

These considerations reinforce the general guidelines that fully implicit schemes are well suited to decay or relaxation processes but should not be used for growth processes, while conversely fully explicit schemes are well suited to growth processes but may require the use of unacceptably small values of Δt to simulate decay processes. Those guidelines are very useful in developing suitable difference schemes for processes involving growth or decay, and they are not as widely appreciated as they deserve to be. We hasten to add, however, that processes involving rapid decay do not by any means constitute the only rationale for the use of implicit schemes. Implicit schemes are also very useful, and sometimes essential, in the numerical approximation of neutrally stable physical phenomena with very small amplitudes but very short physical time scales, for which the use of explicit schemes would entail unacceptably small stability restrictions on Δt. This combination of small amplitudes and short time scales is the hallmark of *stiffness*, which in its many and various forms is the primary motivation for the use of implicit schemes. In fluid dynamics, the most important example of stiffness is provided by sound waves at very low Mach number, which will be discussed in Sect. 8.1.2.

5.2. Nonlinear Source Terms

The source terms encountered in CFD are rarely strictly linear in the dependent variables; they are usually at least weakly nonlinear, and often strongly nonlinear. For simplicity we shall restrict attention to nonlinear source terms that are proportional to some power p of the dependent variable, where $p > 1$. Chemical reactions provide an important example of source terms of this type. The simplest situation is that of a single equation for a single dependent variable f, in which case Eq. (5.1) is replaced by

$$\frac{\partial f}{\partial t} = \beta f^p \tag{5.19}$$

the analytical solution of which is

$$f(t) = [f_0^{1-p} + (1 - p)\beta t]^q \tag{5.20}$$

where $f_0 \equiv f(0)$ and $q \equiv 1/(1 - p)$. Just as in the case of Eq. (5.1), we note that this solution exhibits unbounded growth for $\beta > 0$ and decays to zero for $\beta < 0$, in both cases monotonically. The growth when $\beta > 0$ is faster than exponential, so fast indeed that the solution diverges in a finite time. In realistic situations this divergence is prevented by other terms in the equations. Conversely, the decay when $\beta < 0$ is slower than exponential; in that case $f(t) \sim t^q$ for large t and hence becomes increasingly slower for larger values of p.

We shall rely on our previous experience with linear source terms to limit the present discussion to the particular special cases and numerical schemes of greatest interest. In particular, there is no apparent reason to expect the time-centered Crank–Nicholson scheme to be of any more interest here than it was for Eq. (5.1), so we shall restrict attention to the fully explicit and fully implicit schemes. Moreover, based on our experience as to which schemes are best suited to processes involving growth and decay, we anticipate that the fully explicit and implicit schemes are appropriate for use when $\beta > 0$ and $\beta < 0$, respectively, so we shall further restrict attention to those choices. The use of the fully explicit scheme of Eq. (3.4) to approximate Eq. (5.19) for $\beta > 0$ is straightforward, indeed essentially trivial, and bears a close and obvious resemblance to Eq. (5.3). This case requires no particular elaboration, so we may focus the remainder of our discussion on the numerical approximation of Eq. (5.19) for $\beta < 0$ using the fully implicit scheme of Eq. (3.5), and linearly implicit schemes derived from it.

First we must deal with another minor notational annoyance: in the present context we now have two types of superscripts which must be carefully distinguished. In particular, we must keep in mind that superscripts n and $n + 1$ denote time levels in the usual way, while other superscripts such as p and q denote exponents. This should cause no confusion as long as we do not allow our minds to wander while we are doing algebra, which of course is inadvisable anyway. In the present context, the difference approximation to f^p at time $t = t^n$ will normally be written as $(f^n)^p$ rather than $(f^p)^n$. However, either form should be clear as long as we simply remember the significance of the superscripts. With this convention, the fully implicit difference approximation to Eq. (5.19) for $\beta < 0$ takes the form

$$\frac{f^{n+1} - f^n}{\Delta t} = -|\beta|(f^{n+1})^p \tag{5.21}$$

and we immediately note that in contrast to the analogous linear scheme of Eq. (5.9), Eq. (5.21) is nonlinear in f^{n+1} and cannot in general be analytically solved for f^{n+1}. In more complicated situations involving systems of equations in several variables containing coupled nonlinear source terms (which is typical of problems involving chemical reactions), one obtains a coupled nonlinear system of equations to be solved for the dependent variables at the advanced time level. Single equations or equation systems of this type must in general be solved by iterative methods, such as the Newton–Raphson method. There are indeed many situations in which this is necessary to obtain physically realistic numerical solutions, and hence simply cannot be avoided. However, there are also many situations in which reasonable numerical solutions can be obtained by further approximating $(f^{n+1})^p$ by an expression that is linear in f^{n+1}. This results in a linearly implicit scheme of the type discussed in Sect. 3.2, which can then be immediately solved for f^{n+1}. In cases involving several nonlinearly coupled dependent variables, the resulting linearly implicit scheme requires the solution of a coupled system of linear equations to determine the dependent variables at the advanced time level. The saving grace is that source terms do not involve spatial derivatives, so there is no direct coupling between neighboring spatial mesh points or cells. The dimensionality of such systems is then simply the number of dependent variables they involve, and hence is relatively small. Consequently, such systems are readily susceptible to direct solution by Gaussian elimination or even Cramer's rule, and normally do not require the use of iterative methods.

We therefore proceed to consider linearly implicit approximations to Eq. (5.21) obtained by consistently approximating $(f^{n+1})^p$ therein by expressions that are linear in f^{n+1}. As mentioned in Sect. 3.2, there are basically two ways to proceed. The first and most obvious is to write $f^{n+1} = f^n + \Delta f$, where $\Delta f \equiv f^{n+1} - f^n$, expand $(f^{n+1})^p$ as a Taylor or binomial series in powers of Δf, and discard all terms of quadratic and higher order in Δf. We thereby obtain

$$
\begin{aligned}
(f^{n+1})^p &\approx (f^n)^p + p(f^n)^{p-1}\Delta f \\
&= (f^n)^p + p(f^n)^{p-1}(f^{n+1} - f^n) \\
&= (1-p)(f^n)^p + p(f^n)^{p-1}f^{n+1}
\end{aligned}
\tag{5.22}
$$

Combining this approximation with Eq. (5.21), we obtain the linearly implicit scheme

$$
\frac{f^{n+1} - f^n}{\Delta t} = -|\beta|[(1-p)(f^n)^p + p(f^n)^{p-1}f^{n+1}]
\tag{5.23}
$$

which can immediately be solved for f^{n+1}, with the result

$$f^{n+1} = \left[\frac{1 + |\beta|\Delta t(p - 1)(f^n)^{p-1}}{1 + |\beta|\Delta tp(f^n)^{p-1}} \right] f^n \qquad (5.24)$$

The second way of constructing a consistent linearized approximation to $(f^{n+1})^p$ is based on the observation that f^n is a consistent first-order accurate approximation to f^{n+1}, from which it readily follows that $(f^n)^{p-1}$ is likewise a consistent first-order accurate approximation to $(f^{n+1})^{p-1}$. But $(f^{n+1})^p = (f^{n+1})^{p-1}f^{n+1}$, so that $(f^n)^{p-1}f^{n+1}$ is a consistent first-order accurate approximation to $(f^{n+1})^p$. Combining this approximation with Eq. (5.21), we obtain the alternative linearly implicit scheme

$$\frac{f^{n+1} - f^n}{\Delta t} = -|\beta|(f^n)^{p-1}f^{n+1} \qquad (5.25)$$

which also can immediately be solved for f^{n+1}, with the result

$$f^{n+1} = \frac{f^n}{1 + |\beta|\Delta t(f^n)^{p-1}} \qquad (5.26)$$

Equations (5.23) and (5.25) therefore represent two alternative consistent linearly implicit approximations to the fully implicit scheme of Eq. (5.21).

A heuristic Fourier stability analysis of Eq. (5.23) or (5.25) would require linearization about an unsteady unperturbed solution f_0^n, which is no longer trivial as it was in the case of linear source terms. As a result, those analyses are remarkably tedious, especially considering that the equations do not even involve spatial derivatives. Moreover, the results are difficult to interpret properly due to subtleties beyond the scope of this discussion. Ambitious readers are encouraged to perform those analyses and explore those subtleties, which are actually rather interesting. For present purposes, however, they would represent an unwarranted digression, so we shall eschew the Fourier stability analyses of Eqs. (5.23) and (5.25) and restrict our attention to their qualitative behavior for large Δt. Inspection of Eq. (5.24) shows that f^{n+1} approaches the nonzero limiting value $(1 - 1/p)f^n$ as $\Delta t \to \infty$, whereas Eq. (5.26) shows that f^{n+1} tends to zero for large Δt, just as the analytical solution Eq. (5.20) does for large t. We therefore conclude that the linearly implicit scheme of Eq. (5.25) better represents the qualitative behavior of the exact solution for large Δt, and is consequently preferable to the alternative linearly implicit scheme of Eq. (5.23). It also has the advantage of being simpler in structure.

The linearly implicit schemes of Eqs. (5.23) and (5.25) are both easily generalized to arbitrary nonlinear source terms, for which Eq. (5.19) is replaced by $\partial f/\partial t = G(f)$, where the nonlinear function $G(f)$ is arbitrary. The resulting schemes are given by

$$\frac{f^{n+1} - f^n}{\Delta t} = G(f^n) + \left(\frac{dG}{df}\right)^n (f^{n+1} - f^n) \tag{5.27}$$

$$\frac{f^{n+1} - f^n}{\Delta t} = \frac{G(f^n)}{f^n} f^{n+1} \tag{5.28}$$

respectively, both of which can easily be solved explicitly for f^{n+1}.

Chapter 6

DIFFUSION

The governing Eq. (1.1) for single-component fluids (i.e., pure fluids composed of a single type or species of atoms or molecules) contains terms representing two different types of diffusional processes: the diffusional transport of momentum and energy. In fluid mixtures a third type of diffusional terms occurs, representing the diffusional transport of the masses of the different components or species in the mixture relative to one another. The *transport coefficients* to which these different types of diffusional processes are proportional are the viscosities, thermal conductivity, and mass diffusivities or diffusion coefficients, respectively. These diffusional processes are physically distinct but are closely related, and they are consequently represented by terms of similar, albeit not identical, form. However, they all have essentially the same mathematical character, which is exemplified by the simplest and most basic form of the diffusion equation in one space dimension:

$$\frac{\partial f}{\partial t} = D\frac{\partial^2 f}{\partial x^2} \tag{6.1}$$

where D is the diffusivity, which is here presumed to be a given constant.

Diffusion is fundamentally an irreversible decay process that represents the relaxation of inhomogeneities in f toward a uniform equilibrium state in which $\partial f/\partial x = 0$. (In cases where this is inconsistent with the boundary conditions, f relaxes toward a nonequilibrium steady state in which the mean-squared value of $\partial f/\partial x$ is a miminum.) Diffusional behavior is consequently somewhat analogous to that of source terms describing decay; i.e., source terms with negative coefficients, or $\beta < 0$ in the notation of

Chapter 5. Indeed, many of the features of the various numerical schemes for source terms discussed in Chapter 5 have close diffusional analogs, as discussed below. As in Chapter 5, the three basic time-differencing schemes that we shall consider for diffusion are the fully explicit, fully implicit, and time-centered Crank–Nicholson schemes, which in the present context respectively take the forms

$$\frac{f_i^{n+1} - f_i^n}{\Delta t} = D \left\langle \frac{\partial^2 f}{\partial x^2} \right\rangle_i^n \tag{6.2}$$

$$\frac{f_i^{n+1} - f_i^n}{\Delta t} = D \left\langle \frac{\partial^2 f}{\partial x^2} \right\rangle_i^{n+1} \tag{6.3}$$

$$\frac{f_i^{n+1} - f_i^n}{\Delta t} = \frac{D}{2} \left[\left\langle \frac{\partial^2 f}{\partial x^2} \right\rangle_i^n + \left\langle \frac{\partial^2 f}{\partial x^2} \right\rangle_i^{n+1} \right] \tag{6.4}$$

In all three cases, we shall approximate $\partial^2 f / \partial x^2$ by the second-order accurate spatial difference approximation Eq. (2.9):

$$\left\langle \frac{\partial^2 f}{\partial x^2} \right\rangle_i^n = \frac{f_{i+1}^n - 2f_i^n + f_{i-1}^n}{\Delta x^2} \tag{6.5}$$

which is symmetrical about the central mesh point i. The resulting schemes are consequently often referred to as the forward-time, centered-space (FTCS), backward-time, centered space (BTCS), and centered-time, centered space (CTCS) schemes, respectively.

6.1. The Fully Explicit or FTCS Scheme

The FTCS scheme for diffusion is obtained by combining Eqs. (6.2) and (6.5), with the result

$$\frac{f_i^{n+1} - f_i^n}{\Delta t} = D \frac{f_{i+1}^n - 2f_i^n + f_{i-1}^n}{\Delta x^2} \tag{6.6}$$

As should already be obvious, this scheme is first-order accurate in time and second-order accurate in space. The next step is to analyze its stability by means of a Fourier stability analyis. Since Eq. (6.6) is already linear and contains no inhomogeneous terms, the perturbation δf_i^n satisfies precisely the same equation, so that

$$\frac{\delta f_i^{n+1} - \delta f_i^n}{\Delta t} = D\frac{\delta f_{i+1}^n - 2\delta f_i^n + \delta f_{i-1}^n}{\Delta x^2} \tag{6.7}$$

Substituting $\delta f_i^n = \hat{f}\xi^n \exp\{\iota\kappa i\Delta x\}$ into Eq. (6.7) and canceling out the common factors of $\hat{f}\xi^n \exp\{\iota\kappa i\Delta x\}$, we obtain

$$\frac{\xi - 1}{\Delta t} = \frac{D}{\Delta x^2}\left(e^{\iota\kappa\Delta x} - 2 + e^{-\iota\kappa\Delta x}\right) = \frac{2D}{\Delta x^2}[\cos(\kappa\Delta x) - 1] \tag{6.8}$$

so that

$$\xi(\kappa) = 1 + \frac{2D\Delta t}{\Delta x^2}[\cos(\kappa\Delta x) - 1] \tag{6.9}$$

Since ξ is real, the stability condition of Eq. (4.4) reduces to $-1 \le \xi(\kappa) \le 1$ for all κ. But $\cos(\kappa\Delta x) - 1$ is never positive, so the inequality $\xi(\kappa) \le 1$ is always satisfied. The stability condition then reduces to $\xi(\kappa) \ge -1$, or

$$\frac{D\Delta t}{\Delta x^2} \le \frac{1}{1 - \cos(\kappa\Delta x)} \tag{6.10}$$

for all κ. The right member of this equation is smallest when $\cos(\kappa\Delta x) = -1$, corresponding to the shortest allowed wavelength or largest allowed wavenumber of $\kappa = \pi/\Delta x$. The overall stability condition then becomes $D\Delta t/\Delta x^2 \le 1/2$, or

$$\Delta t \le \frac{\Delta x^2}{2D} \tag{6.11}$$

Exercise 6.1. (a) Show that in two and three space dimensions, Δx^2 in Eq. (6.11) is replaced by $(1/\Delta x^2 + 1/\Delta y^2)^{-1}$ and $(1/\Delta x^2 + 1/\Delta y^2 + 1/\Delta z^2)^{-1}$, respectively. (b) Show that these results can be interpreted as special cases of Eq. (4.12).

A noteworthy feature of Eq. (6.11) is that the maximum permissible Δt is quadratic in Δx, which implies that as the mesh is refined (i.e., as $\Delta x \to 0$), Δt goes to zero much faster than Δx. Thus doubling the spatial resolution (reducing Δx by a factor of two) requires reducing Δt by a factor of four, reducing Δx by a factor of 10 requires reducing Δt by a factor of 100, and so on. This behavior often makes the FTCS scheme unsuitable in problems where the diffusional or viscous terms are comparable to or larger than the other terms in the equations; e.g., flow at low Reynolds number. The resulting stringent restrictions on the time step are particularly frustrating in problems where the diffusional terms are only important in localized small regions of the computing mesh, such as boundary layers. The above considerations further imply that in explicit calculations involving equations with diffusional terms, repeatedly reducing Δt and Δx by the same factor will eventually result in instability. This is useful to remember, as it suggests that instabilities encountered in that manner are likely to be diffusional in nature.

It is of course necessary for Δt to satisfy Eq. (6.11) so that the calculation is stable, but as previously discussed in connection with source terms, stability alone is not sufficient to ensure that the resulting numerical solution manifests qualitatively reasonable physical behavior. The partial differential diffusion Eq. (6.1) predicts that an individual continuous Fourier mode of the form $f_0 \exp\{\iota(\kappa x - \omega t)\}$ evolves in time according to the *dispersion relation* $\omega = -\iota\kappa^2 D$, so that the time factor $\exp\{-\iota\omega t\} = \exp\{-\kappa^2 D t\}$ decays exponentially to zero with a time constant of $(\kappa^2 D)^{-1}$. In order for the numerical scheme to exhibit qualitatively similar behavior, $\xi^n = \exp\{-\iota\omega n\Delta t\}$ must decay monotonically to zero with increasing n. This in turn requires $\xi(\kappa)$ to remain positive, for otherwise the solution would change sign on alternate time steps and the decay would be oscillatory rather than monotonic. In order to prevent such qualitatively unphysical behavior, we must therefore impose the condition $\xi(\kappa) \geq 0$, which is more restrictive than the condition $\xi(\kappa) \geq -1$ required by stability alone. Combining the condition $\xi(\kappa) \geq 0$ with Eq. (6.9), we obtain

$$\frac{2D\Delta t}{\Delta x^2} \leq \frac{1}{1 - \cos(\kappa\Delta x)} \tag{6.12}$$

which again is most restrictive when $\cos(\kappa\Delta x) = -1$, when it reduces to

$$\Delta t \leq \frac{\Delta x^2}{4D} \tag{6.13}$$

This condition is more restrictive than the stability condition of Eq. (6.11) by a factor of two. Thus a safety factor of $1/2$ must be inserted into the stability condition to prevent the numerical solution from producing qualitatively unphysical behavior.

The highly restrictive stability condition of the FTCS scheme provides considerable incentive to explore the use of alternative implicit schemes instead. The simplest, and in many respects the best, scheme of this type is the fully implicit or BTCS scheme, which we now proceed to discuss.

6.2. The Fully Implicit or BTCS Scheme

The fully implicit scheme for diffusion is obtained by combining Eqs. (6.3) and (6.5), with the result

$$\frac{f_i^{n+1} - f_i^n}{\Delta t} = D \frac{f_{i+1}^{n+1} - 2f_i^{n+1} + f_{i-1}^{n+1}}{\Delta x^2} \tag{6.14}$$

This scheme too is obviously first-order accurate in time and second-order accurate in space. It requires the solution of a coupled linear system of equations to determine the quantities f_i^{n+1}. Those equations are tridiagonal in form and can easily be directly solved by means of the Thomas algorithm, as discussed in Sect. 3.3. The stability properties of the fully implicit scheme can be determined by the same procedure used in the previous section, and we thereby obtain

$$\frac{\xi - 1}{\Delta t} = \frac{\xi D}{\Delta x^2} \left(e^{\iota \kappa \Delta x} - 2 + e^{-\iota \kappa \Delta x} \right) = \frac{2\xi D}{\Delta x^2} [\cos(\kappa \Delta x) - 1] \tag{6.15}$$

so that

$$\xi(\kappa) = \left\{ 1 + \frac{2D\Delta t}{\Delta x^2} [1 - \cos(\kappa \Delta x)] \right\}^{-1} \tag{6.16}$$

Since $1 - \cos(\kappa \Delta x) \geq 0$, we see that $0 \leq \xi \leq 1$ for all values of κ and Δt, so that the BTCS scheme is unconditionally stable. Moreover, it automatically exhibits the desired qualitatively physical monotonic decay behavior discussed in the previous section, again for all Δt.

Exercise 6.2. (a) Show that the difference scheme obtained by combining Eq. (6.3) with Eq. (2.20) instead of Eq. (6.5) is algebraically equivalent to the FTCS scheme of Eq. (6.6) with Δt replaced by $(1/\Delta t + 2D/\Delta x^2)^{-1}$. (b) Show that the resulting numerical solution f_i^n consequently approximates the true solution at the time $t = n\Delta t(1 + 2D\Delta t/\Delta x^2)^{-1}$ rather than $t = n\Delta t$ as it should. Thus the inconsistency of Eq. (2.20) manifests itself in this context as an unphysical distortion of the time scale.

6.3. The Crank–Nicholson or CTCS Scheme

The Crank–Nicholson scheme for diffusion is obtained by combining Eqs. (6.4) and (6.5), with the result

$$\frac{f_i^{n+1} - f_i^n}{\Delta t} = \frac{D}{2} \left(\frac{f_{i+1}^n - 2f_i^n + f_{i-1}^n}{\Delta x^2} + \frac{f_{i+1}^{n+1} - 2f_i^{n+1} + f_{i-1}^{n+1}}{\Delta x^2} \right) \tag{6.17}$$

In contrast to the fully explicit and fully implicit schemes, the Crank–Nicholson scheme is second-order accurate in time as well as space. Like the fully implicit scheme, it requires the solution of a tridiagonal linear equation system to determine the quantities f_i^{n+1}, which again can easily be accomplished using the Thomas algorithm. The stability analysis proceeds in the usual manner, and yields

$$\frac{\xi - 1}{\Delta t} = (\xi + 1)\frac{D}{\Delta x^2}[\cos(\kappa\Delta x) - 1] \tag{6.18}$$

Solving for the growth factor ξ, we obtain

$$\xi(\kappa) = \frac{1 - (D\Delta t/\Delta x^2)[1 - \cos(\kappa\Delta x)]}{1 + (D\Delta t/\Delta x^2)[1 - \cos(\kappa\Delta x)]} \tag{6.19}$$

Since ξ is real, the stability condition of Eq. (4.4) again reduces to $-1 \leq \xi(\kappa) \leq 1$ for all κ, and it is easy to show that both of those inequalities are satisfied for all Δt.

Exercise 6.3. Verify the previous statement.

The Crank–Nicholson scheme for diffusion is therefore unconditionally stable, and this property together with its second-order accuracy has made it very popular. In contrast to the fully implicit scheme, however, the Crank–Nicholson scheme does not unconditionally satisfy the condition $\xi(\kappa) \geq 0$ required to prevent qualitatively unphysical solution behavior, as discussed in Sect. 6.1. The latter condition combines with Eq. (6.19) to become

$$\frac{D\Delta t}{\Delta x^2}[1 - \cos(\kappa\Delta x)] \leq 1 \tag{6.20}$$

Once again, this condition is most restrictive when $\cos(\kappa\Delta x) = -1$, when it reduces to $2D\Delta t/\Delta x^2 \leq 1$. But this is precisely equivalent to the explicit stability condition of Eq. (6.11)! The unconditionally stable Crank–Nicholson scheme consequently exhibits qualitatively unphysical behavior when Δt exceeds the stability limit for the fully explicit scheme. This behavior often manifests itself as bounded unphysical parasitic oscillations that reverse sign on every time step. Such oscillations have been observed and discussed by numerous authors, but curiously do not seem to have greatly diminished the popularity of the scheme.

In summary, both the fully implicit and the Crank–Nicholson schemes are unconditionally stable, and hence both allow the user to exceed the explicit stability condition of Eq. (6.11) by sacrificing some of the accuracy of the simulation. Moreover, both schemes require the solution of a linear equation system of the same structure in order to determine the advanced-time variables f_i^{n+1} on each time step. The essential differences between the two schemes are that the fully implicit scheme is first-order accurate in time but exhibits qualitatively reasonable solution behavior for all Δt, while the Crank–Nicholson scheme is second-order accurate in time but exhibits qualitatively unphysical behavior when Δt exceeds the explicit stability condition of Eq. (6.11). Moreover, as discussed in Sect. 2.4, the second-order accuracy in time only translates into improved solution accuracy for relatively small time steps; i.e., values of Δt of the same order of magnitude as the explicit stability limit. Even in that case, similar accuracy improvements could be obtained more easily, and probably more economically as well, simply by using the fully explicit scheme with a smaller value of Δt. The fully implicit scheme has the further advantage of producing the steady-state solution in a single time step when Δt is very large. Thus, although its second-order accuracy makes the Crank–Nicholson scheme superficially attractive, under closer scrutiny its appeal is found to be specious. For small Δt, it is inferior to the much simpler fully explicit scheme, which delivers the values of f_i^{n+1} directly to the doorstep without requiring the solution of a linear algebraic equation system. For larger Δt, the Crank–Nicholson scheme is inferior to the fully implicit scheme for the reasons discussed above. These undesirable features are closely analogous to those previously found for decaying source terms, as discussed in Chapter 5. The verdict seems clear: the Crank–Nicholson scheme is of little if any interest for diffusion problems, although this assessment is by no means unanimous among numerical analysts.

6.4. Alternating-Direction Implicit Schemes

The diffusion equation provides a convenient context in which to familiarize the reader with the basic ideas of alternating-direction implicit (ADI) schemes, although such schemes are of much wider applicability. ADI schemes constitute a large extended family of computationally efficient partially implicit schemes for problems in two or three space dimensions. These schemes are designed to exploit the Thomas tridiagonal solution algorithm mentioned in Sect. 3.3, which serves as both their *raison d'etre* and *sine qua non*.

The rationale of ADI schemes is to construct difference equations that are implicit in only one direction *at a time* (i.e., on any particular time step), and to alternate the unidirectional implicitness between or among the different spatial directions on successive time steps. This process will be illustrated within the context of diffusion in two space dimensions (x, y), in which Eq. (6.1) is replaced by

$$\frac{\partial f}{\partial t} = D \left(\frac{\partial^2 f}{\partial x^2} + \frac{\partial^2 f}{\partial y^2} \right) \tag{6.21}$$

Let us consider the numerical solution of Eq. (6.21) using difference schemes of the general form

$$\frac{f_{ij}^{n+1} - f_{ij}^n}{\Delta t} = D \left[\left\langle \frac{\partial^2 f}{\partial x^2} \right\rangle_{ij}^p + \left\langle \frac{\partial^2 f}{\partial y^2} \right\rangle_{ij}^q \right] \tag{6.22}$$

where f_{ij}^n is the finite-difference approximation to $f(x, y, t)$ at the space-time point $(x, y, t) = (i\Delta x, j\Delta y, n\Delta t)$, and the spatial difference approximations $\langle \partial^2 f / \partial x^2 \rangle_{ij}^p$ and $\langle \partial^2 f / \partial y^2 \rangle_{ij}^q$ are defined by obvious minor modifications of Eq. (6.5). This scheme includes the basic fully explicit and fully implicit schemes as the special cases $p = q = n$ and $p = q = n + 1$, respectively, of which our interest here centers on the latter.

Exercise 6.4. Verify that like its one-dimensional analog Eq. (6.14), the fully implicit scheme of Eq. (6.22) with $p = q = n + 1$ is unconditionally stable and exhibits the desired qualitative solution behavior for all Δt.

Equation (6.22) with $p = q = n + 1$ constitutes a large sparse system of linear equations that must be solved on each time step to determine the quantities f_{ij}^{n+1}. But unlike its one-dimensional analog, this system is not tridiagonal in structure, since it simultaneously involves the values of f_{ij}^{n+1} at the point (i, j) and all of its immediate neighbors $(i \pm 1, j)$ and $(i, j \pm 1)$ in both the x-and y-directions. As a result, it cannot be solved by means of the Thomas algorithm. The corresponding ADI scheme can be thought of as a further approximation to the fully implicit scheme, in which enough implicitness is retained to preserve the unconditional stability while enough is removed to restore the tridiagonal structure of the linear equations that must be solved to determine f_{ij}^{n+1} on each time step, so that the Thomas algorithm can again be employed. This is accomplished simply by setting $(p, q) = (n + 1, n)$ when n is odd and $(p, q) = (n, n + 1)$ when n is even, so that the implicitness alternates in direction on successive time steps.

Exercise 6.5. Show that this procedure has the further beneficial side effect of rendering the resulting scheme second-order accurate in time over a period of two time steps.

Analyzing the stability of the resulting ADI scheme requires a slight modification of the basic Fourier method, since the difference equations have a different form, and hence a different amplification factor, on successive time steps. Upon inspection, however, it becomes clear that the effective amplification factor for the scheme over a period of two time steps is simply the product of their separate amplification factors. This can also be seen by restricting attention to either odd or even values of n and regarding the ADI scheme as a whole as a two-step composite scheme for advancing the numerical solution from time level n to time level $n + 2$. It is then straightforward to show that the ADI scheme is unconditionally stable, but unfortunately it does not preserve the desirable qualitative properties of the fully implicit scheme for large Δt. Indeed, it exhibits the same type of qualitatively unphysical solution behavior as the Crank–Nicholson scheme in one space dimension, which tends to produce irregular or oscillatory solutions when Δt significantly exceeds the explicit stability limit. In the present context, it also exhibits a second type of unphysical behavior due to the artificial anisotropy introduced by preferential transport in the x- and y-directions. For large values of Δt, this can result in the development of artificial protrusions parallel to the x and y axes (the so-called starfish effect). To prevent these various unphysical solution features from developing, it is once again necessary to restrict Δt to values comparable to the explicit stability limit. This restriction significantly limits the practical utility of time-marching ADI schemes for unsteady flow calculations. It should be noted, however, that ADI schemes are also advantageous in accelerating the convergence of linear iterative methods such as SOR, and in that context the above drawbacks do not apply.

Chapter 7

CONVECTION

The primary dependent variables in fluid dynamics are the local densities of the globally conserved quantities mass, momentum, and energy. All of those densities are subject to convection; i.e., they are transported from one location to another by the motion of the fluid, so that in general their local values at a fixed point in space will vary with time as different Lagrangian fluid particles with different densities pass through that point. The time evolution equations for those densities consequently all contain convection terms representing their transport due to fluid motion. All such terms have essentially the same origin and mathematical character, and consequently can be treated by essentially the same numerical methods.

First a word about terminology. There is considerable ambiguity and lack of unanimity in the literature as to whether the process described above should be referred to as "convection" or "advection." No attempt has been made to trace the etymology of those terms, but the latter seems to be of more recent origin, and may have been coined to distinguish passive transport by a velocity field, which is an essentially kinematic process, from the more dynamical processes that have traditionally been referred to as natural and forced convection. In many contexts the two terms are virtually synonymous, and we shall regard them as such, but with a bias in favor of "convection." However, the reader is cautioned that different authors sometimes invest them with more specialized meanings.

As will be seen below, stable explicit schemes for convection are typically subject to stability conditions of the form $|u|\Delta t/\Delta x \leq A$, where u is the fluid velocity and A is a scheme-dependent constant of order unity. Such conditions have the physical interpretation that the distance the fluid travels on a single time step must be comparable to or somewhat less than the mesh spacing. Conditions of this type are called *Courant-Friedrichs-Lewy* (CFL) *conditions*, or *Courant conditions* for short, and the dimensionless ratio $|u|\Delta t/\Delta x$ is often referred to as the *Courant number*. Such conditions are not peculiar to convection, but are a generic feature of explicit schemes for hyperbolic partial differential equations, with $|u|$ replaced by the associated finite signal propagation speed. We shall see in the next chapter that stable explicit schemes for sound waves are accordingly subject to similar CFL conditions with $|u|$ replaced by the sound speed c. More generally, when convection and sound waves are considered simultaneously, the CFL condition typically involves $|u| + c$, as one would intuitively expect from characteristics considerations.

In most unsteady or transient fluid dynamics problems, the convective CFL condition is found to be necessary for solution accuracy as well as stability. When this is the case, there is no advantage to be gained by using implicit schemes to alleviate the explicit stability condition, since the CFL condition would still remain in effect as an accuracy condition. For this reason, the present discussion will be restricted to explicit convective schemes. It should be noted, however, that occasional exceptions occur in which the use of considerably larger time steps would not entail a significant sacrifice in accuracy, and implicit convective schemes can then be highly advantageous. Such situations typically occur when the explicit stability limit is controlled by relatively high fluid velocities and/or relatively small mesh spacings in localized regions where the flow is nearly steady, while the velocities are significantly lower and/or the mesh spacings are significantly larger in the remaining unsteady portions of the flow field, where considerably larger time steps could otherwise be employed. An unconditionally stable implicit scheme would then allow the use of larger time steps in such situations, but such schemes will not be discussed here.

The simplest equation describing convection or advection in one space dimension is

$$\frac{\partial f}{\partial t} + u\frac{\partial f}{\partial x} = 0 \tag{7.1}$$

where the velocity u is a constant independent of x and t. It is easy to verify that the exact general solution of this equation is given by

$$f(x, t) = f_0(x - ut) \tag{7.2}$$

where $f_0(x) = f(x, 0)$ is the initial condition. This solution simply represents the fact that the initial profile $f_0(x)$ is shifted or translated by a distance ut at time t, but with no change in its shape. This exact solution provides a useful basis for comparison with approximate numerical solutions.

A generic fully explicit finite-difference scheme for Eq. (7.1) is given by

$$\frac{f_i^{n+1} - f_i^n}{\Delta t} = -u \left\langle \frac{\partial f}{\partial x} \right\rangle_i^n \tag{7.3}$$

in which the spatial difference approximation represented by $\langle \partial f / \partial x \rangle_i^n$ remains to be specified. We now proceed to consider several special cases of Eq. (7.3) that result from the use of particular spatial difference approximations.

7.1. Centered Spatial Differencing

We first consider the symmetrical or centered spatial difference approximation of Eq. (2.5):

$$\left\langle \frac{\partial f}{\partial x} \right\rangle_i^n = \frac{f_{i+1}^n - f_{i-1}^n}{2\Delta x} \tag{7.4}$$

Combining Eqs. (7.3) and (7.4), we obtain the basic FTCS scheme for convection:

$$\frac{f_i^{n+1} - f_i^n}{\Delta t} = -u \left(\frac{f_{i+1}^n - f_{i-1}^n}{2\Delta x} \right) \tag{7.5}$$

This scheme is obviously first-order accurate in time and second-order accurate in space, just like the analogous FTCS scheme for diffusion. The Fourier stability analysis of Eq. (7.5) again proceeds in the usual manner, and we thereby obtain

$$\frac{\xi - 1}{\Delta t} = -u \left(\frac{e^{\iota \kappa \Delta x} - e^{-\iota \kappa \Delta x}}{2\Delta x} \right) = -u \left[\frac{\iota \sin(\kappa \Delta x)}{\Delta x} \right] \tag{7.6}$$

Solving for ξ, we encounter our first complex amplification factor:

$$\xi(\kappa) = 1 - \iota \left(\frac{u\Delta t}{\Delta x} \right) \sin(\kappa \Delta x) \tag{7.7}$$

so that

$$|\xi(\kappa)|^2 = 1 + \left(\frac{u\Delta t}{\Delta x} \right)^2 \sin^2(\kappa \Delta x) \tag{7.8}$$

We see that $|\xi(\kappa)| \geq 1$ for all κ and Δt, so that the stability condition of Eq. (4.4) cannot be satisfied for any nonzero value of Δt, however small. In contrast to the very useful FTCS scheme for diffusion, the analogous FTCS scheme for convection is therefore unconditionally unstable and hence entirely useless.

Exercise 7.1. Show that the corresponding fully implicit BTCS scheme for convection is unconditionally stable.

Why is the same type of scheme conditionally stable and useful for diffusion, but not for convection? The fundamental answer to this question is simply that such are the vicissitudes of life. However, this does not imply that this difference in behavior is incomprehensible. The unconditional instability of Eq. (7.5) can be qualitatively and intuitively understood in terms of its truncation errors. For sufficiently small values of Δx and Δt, first-order truncation errors would be expected to dominate those of higher order, so it is logical to begin by examining the form of the first-order errors, and then if necessary to proceed to the second-order errors, and so on. Since Eq. (7.5) is second-order accurate in space, the only first-order truncation error is that arising from the first-order accurate forward difference approximation to the time derivative:

$$\frac{f_i^{n+1} - f_i^n}{\Delta t} = \left(\frac{\partial f}{\partial t}\right)_i^n + \frac{1}{2}\Delta t \left(\frac{\partial^2 f}{\partial t^2}\right)_i^n + \mathcal{O}(\Delta t^2) \tag{7.9}$$

In analyses of this type, it is usually advantageous to transform second and higher time derivatives appearing in truncation errors into spatial derivatives. The reason is that the resulting modified differential equations, to which the difference scheme is equivalent, are then first-order in time, just like the true physical time evolution equations, and this usually reveals their physical behavior more clearly. As discussed in Sect. 3.2, this transformation must be performed by manipulating the modified differential equation equivalent to Eq. (7.5) rather than the original differential Eq. (7.1). Since Eq. (7.5) is first-order accurate in time and second-order accurate in space, the modified equation has the form

$$\frac{\partial f}{\partial t} + u\frac{\partial f}{\partial x} = \mathcal{O}(\Delta t) + \mathcal{O}(\Delta x^2) \tag{7.10}$$

Differentiating Eq. (7.10) with respect to time and combining the result with Eq. (7.10) itself, we obtain

$$\frac{\partial^2 f}{\partial t^2} = -u \left(\frac{\partial^2 f}{\partial t \partial x} \right) + \mathcal{O}(\Delta t) + \mathcal{O}(\Delta x^2)$$

$$= u^2 \left(\frac{\partial^2 f}{\partial x^2} \right) + \mathcal{O}(\Delta t) + \mathcal{O}(\Delta x^2) \tag{7.11}$$

Combining the FTCS scheme of Eq. (7.5) with Eqs. (2.5), (7.9), and (7.11), we find

$$\left(\frac{\partial f}{\partial t} \right)_i^n + u \left(\frac{\partial f}{\partial x} \right)_i^n = -\frac{1}{2} \Delta t u^2 \left(\frac{\partial^2 f}{\partial x^2} \right)_i^n + \mathcal{O}(\Delta t^2) + \mathcal{O}(\Delta x^2) \tag{7.12}$$

which can be regarded as the modified differential equation that the finite-difference scheme of Eq. (7.5) is effectively solving. Thus we see that the first-order truncation error arising from the forward time differencing in Eq. (7.5) effectively introduces an artificial diffusion term into Eq. (7.1), and moreover that the artificial diffusivity in that term has the negative value $D_a = -\frac{1}{2}\Delta t u^2$. Negative diffusion is inherently and explosively unstable, which follows immediately from the diffusional dispersion relation $\omega = -\iota\kappa^2 D$ discussed in Chapter 6. The corresponding time factor is again $\exp\{-\iota\omega t\} = \exp\{-\kappa^2 D t\}$, and when $D < 0$ that factor diverges exponentially with a time constant $(\kappa^2|D|)^{-1}$. Even worse, the growth rate is unbounded with respect to κ and becomes infinite as $\kappa \to \infty$ (i.e., as the wavelength becomes infinitely short). The same pathological behavior occurs when $D > 0$ and $t < 0$, and is the reason why solving the diffusion equation backward in time is an ill-posed problem. This provides a classic illustration of the phenomenon mentioned in the Preface: a numerical ailment with a clear physical interpretation.

The above analysis also provides some valuable insight into how the FTCS scheme of Eq. (7.5) might be modified to remove the unconditional instability. It is clear that the sign of the first-order truncation error would be reversed by the use of backward time differencing, and that this would in turn reverse the sign of the artificial diffusivity D_a, thereby making it positive. This would result in decay instead of growth, which should eliminate the instability. However, the use of backward time differencing would result in a fully implicit scheme, which we would prefer to avoid for the reasons discussed in Sect. 3.2. Fortunately, a similar effect can be obtained in an explicit manner by exploiting the observation that in the context of convection, looking backward in time is like looking upstream in space; i.e., in the opposite direction of u. This suggests that if we were to define the spatial difference approximation $\langle \partial f / \partial x \rangle_i^n$ in terms of the prevailing conditions upstream of the point i, the effect would be similar to that of evaluating it at the point i but at a later time, which in turn would approximate the use of backward time differencing. We are thereby led to consider the important concept of *upwind differencing* of convection terms and equations, which in its many and various forms permeates essentially all of CFD.

7.2. Upwind Spatial Differencing

Upwind differencing is also often referred to as upstream or *donor-cell* differencing. The latter terminology refers to the fact that upstream cells convectively "donate" their contents to downstream (or "acceptor") cells. The present discussion will be limited to upwind differencing in its simplest and most basic forms, but this is sufficient to grasp the essential ideas. For simplicity we shall henceforth presume that $u > 0$, but the modifications required when $u < 0$ are obvious. In real applications, the sign of u is not normally known *a priori*, and this requires suitable logic or testing to determine the upstream direction, but that logic is straightforward. Logical tests can be avoided by the use of equivalent expressions involving sign functions or absolute values, in particular $u / |u|$.

Since $u > 0$ in the present context, the upstream direction is to the left. The simplest way to obtain upwind differencing is then to use the first-order accurate spatial difference approximation of Eq. (2.4) instead of Eq. (2.5), so that

$$\left\langle \frac{\partial f}{\partial x} \right\rangle_i^n = \frac{f_i^n - f_{i-1}^n}{\Delta x} \tag{7.13}$$

Combining Eqs. (7.3) and (7.13), we obtain

$$\frac{f_i^{n+1} - f_i^n}{\Delta t} = -u \left(\frac{f_i^n - f_{i-1}^n}{\Delta x} \right) \tag{7.14}$$

which is now obviously first-order accurate in both space and time. A routine Fourier stability analysis then yields

$$\frac{\xi - 1}{\Delta t} = -u \left(\frac{1 - e^{-\iota \kappa \Delta x}}{\Delta x} \right) \tag{7.15}$$

so that

$$\xi(\kappa) = 1 - \left(\frac{u\Delta t}{\Delta x}\right)\left(1 - e^{-\iota\kappa\Delta x}\right) \tag{7.16}$$

After a little complex algebra, we find

$$|\xi(\kappa)|^2 = 1 - 2C(1 - C)[1 - \cos(\kappa\Delta x)] \tag{7.17}$$

where $C \equiv u\Delta t/\Delta x$ is the convective Courant number. The stability condition of Eq. (4.4) now readily reduces to

$$(1 - C)[1 - \cos(\kappa\Delta x)] \geq 0 \text{ for all } \kappa \tag{7.18}$$

and since $1 - \cos(\kappa\Delta x)$ is never negative, this further reduces to $1 - C \geq 0$, or

$$C = \left(\frac{u\Delta t}{\Delta x}\right) \leq 1 \tag{7.19}$$

The upwind convective scheme of Eq. (7.14) is therefore conditionally stable provided that $\Delta t \leq \Delta x/u$.

It is of interest to examine the first-order truncation errors of Eq. (7.14). Expanding f_{i-1}^n in a Taylor series about the point i, we readily find

$$\frac{f_i^n - f_{i-1}^n}{\Delta x} = \left(\frac{\partial f}{\partial x}\right)_i^n - \frac{1}{2}\Delta x\left(\frac{\partial^2 f}{\partial x^2}\right)_i^n + \mathcal{O}(\Delta x^2) \tag{7.20}$$

Combining Eqs. (7.9), (7.11) and (7.20) with Eq. (7.14), we obtain

$$\left(\frac{\partial f}{\partial t}\right)_i^n + u\left(\frac{\partial f}{\partial x}\right)_i^n = \frac{1}{2}u\Delta x(1 - C)\left(\frac{\partial^2 f}{\partial x^2}\right)_i^n$$
$$+ \mathcal{O}(\Delta x^2) + \mathcal{O}(\Delta t^2) \tag{7.21}$$

which again can be regarded as the modified differential equation that the difference scheme of Eq. (7.14) is effectively solving. Once again the first-order truncation errors contribute a term of diffusional form, in which the artificial diffusivity now has the value $D_a = \frac{1}{2}u\Delta x(1 - C)$, which is nonnegative as long as the stability condition $C \leq 1$ is satisfied. This coefficient would vanish if the calculation could be run with $C = 1$, but it is not feasible to run right at the stability limit, where there is no margin for error. As discussed in Chapter 4, the time step is typically not allowed to exceed half the theoretical stability limit, so that in practice the artificial diffusivity of the upwind difference scheme is generally on the order of $u\Delta x$. This is typically many orders of magnitude larger than physical molecular diffusivities, and it has the undesirable effect of artificially smoothing steep gradients and smearing out the sharp features of the numerical solution. These effects are similar to those of a physical diffusion term with the value $D = D_a$, but the two are not precisely equivalent because of the higher-order truncation errors. This artificial diffusion is the primary disadvantage of upwind differencing, and it is a very serious one. Indeed, in most situations it degrades the solution accuracy to an unacceptable degree, and as a result the basic first-order upwind convective difference scheme of Eq. (7.14) is widely regarded as highly unsatisfactory and unworthy of serious consideration. However, it is not quite so bad as its reputation, and there are many problems in which upwind differencing is more than adequate to compute solutions that are sufficiently accurate for the purpose at hand. Nevertheless, such situations are exceptional, so one is led to seek alternative convective schemes with smaller artificial diffusivities, or preferably none at all. The simplest such schemes are partial or interpolated upwinding schemes, which are discussed in the next section.

The artificial diffusion discussed above is also commonly referred to as "numerical viscosity." This terminology is unfortunate and misleading in two respects. First, viscosity is the transport coefficient associated with the diffusion of fluid momentum or velocity, whereas the artificial diffusion produced by upwind convective differencing affects all dependent variables whose time evolution equations contain convective terms, including the mass and energy densities. Second, in contrast to physical diffusion and viscosity, artificial diffusion does not act isotropically in two and three space dimensions. The artificial transport is primarily in the streamwise direction (i.e., parallel to the fluid velocity vector), and consequently is often insignificant in directions normal or transverse to the flow. In particular, the artificial diffusion due to upwind differencing does not produce artificial boundary layers in parallel flow past solid surfaces, as a physical viscosity would do.

7.3. Partial or Interpolated Upwinding Schemes

The preceding analysis shows that centered convective differencing introduces a negative artificial diffusivity, which produces instability, whereas upwind convective differencing introduces a positive artificial diffusivity, which produces artificial smoothing. One is naturally led to wonder whether the two could be combined in such a way as to eliminate the artificial diffusion altogether. This can indeed be done, and in fact is straightforward. Perhaps the simplest way to proceed is to define $\langle \partial f / \partial x \rangle_i^n$ as a weighted average of the centered and upwind difference expressions given in Eqs. (7.4) and (7.13), and to determine the associated weighting factor by requiring the resulting net artificial diffusivity D_a to vanish. Thus we write

$$\left\langle \frac{\partial f}{\partial x} \right\rangle_i^n = \chi \frac{f_i^n - f_{i-1}^n}{\Delta x} + (1 - \chi) \frac{f_{i+1}^n - f_{i-1}^n}{2 \Delta x}$$

$$= \frac{(1 - \chi) f_{i+1}^n + 2 \chi f_i^n - (1 + \chi) f_{i-1}^n}{2 \Delta x} \qquad (7.22)$$

which reduces to pure upwind differencing when $\chi = 1$ and pure centered differencing when $\chi = 0$. Combining Eqs. (7.3) and (7.22), we obtain

$$\frac{f_i^{n+1} - f_i^n}{\Delta t} = -u \left[\frac{(1 - \chi) f_{i+1}^n + 2 \chi f_i^n - (1 + \chi) f_{i-1}^n}{2 \Delta x} \right] \qquad (7.23)$$

Another standard truncation-error analysis shows that Eq. (7.23) is equivalent to the modified differential equation

$$\left(\frac{\partial f}{\partial t} \right)_i^n + u \left(\frac{\partial f}{\partial x} \right)_i^n = D_a \left(\frac{\partial^2 f}{\partial x^2} \right)_i^n + \mathcal{O}(\Delta x^2) + \mathcal{O}(\Delta t^2) \qquad (7.24)$$

where $D_a = \frac{1}{2} u \Delta x (\chi - C)$. It follows that setting $\chi = C$ in Eq. (7.23) makes $D_a = 0$ and thereby results in a second-order accurate explicit convective difference scheme with zero artificial diffusivity. This scheme is commonly referred to as *second-order upwind* or *interpolated donor cell* differencing. Of course, the elimination of artificial diffusion does not imply the removal of other types of truncation errors, which remain present. In particular, the second-order upwind scheme is subject to *dispersion errors*, which themselves can be quite serious and will be discussed in the next section.

An alternative route to, or interpretation of, the second-order upwind scheme is provided by the following considerations. As discussed in Sect. 7.1, the negative artificial diffusivity and consequent instability of the FTCS convective scheme of Eq. (7.5) is due to the first-order temporal truncation error associated with the forward time differencing. That error would evidently be eliminated by the use of centered time differencing; i.e., if $\langle \partial f / \partial x \rangle$ were evaluated at the point i and time level $n+$ instead of time level n. But the discussion of Sect. 7.1 further suggests that this should be functionally equivalent to evaluating $\langle \partial f / \partial x \rangle$ at time level n but a distance $\frac{1}{2} u \Delta t$ upstream of the point x_i, provided the latter approximation is second-order accurate in space to avoid the appearance of other first-order truncation errors.

Exercise 7.2. Verify that when $\chi = C$, the quantity in square brackets in Eq. (7.23) is indeed a second-order accurate spatial difference approximation to $\partial f/\partial x$ at the space-time point $(x, t) = (x_i - \frac{1}{2}u\Delta t, t^n)$. This shows that the right member of Eq. (7.23) can be interpreted as an upstream spatial interpolation that approximates the effect of centered time differencing.

Exercise 7.3. Show that the second-order upwind scheme is conditionally stable provided that the stability condition $C \leq 1$ is satisfied.

It should be noted, however, that contrary to what one might intuitively expect, the obvious generalization of the second-order upwind scheme to two and three space dimensions is unconditionally unstable due to the presence of destabilizing truncation errors that involve cross derivatives and thus have no one-dimensional analogs.

Exercise 7.4. Confirm the previous statement in a two-dimensional rectangular mesh by means of Fourier stability and truncation-error analyses.

This instability has sometimes been overlooked because it tends to be relatively weak and is localized in regions where the flow is primarily along cell diagonals. As a result, it does not always manifest itself in practical CFD calculations, where it can easily be masked or overshadowed by other physical or numerical dissipative mechanisms. But such fortuity is unreliable, and interpolated donor cell differencing should therefore not be used in two or three space dimensions without appropriate modification. Fortunately, the required modifications are straightforward, as discussed by Dukowicz & Ramshaw, *J. Comput. Phys.* **32**, 71 (1979).

7.4. Dispersion Errors

Dispersion errors are an affliction visited upon those who have the temerity to attempt to mitigate diffusional truncation errors. The essence of dispersion is that Fourier components of the numerical solution with different wavelengths propagate with different speeds, in contrast to the behavior of Eq. (7.2), which shows that all features of the exact solution propagate with the same speed u. This causes the different Fourier modes to artificially drift apart, thereby producing artificial bounded spatial oscillations (or "wiggles") in the numerical solution. Such oscillations tend to be largest in regions of steep gradients, and are frequently large enough to produce an unacceptable degradation in both the accuracy and qualitative features of the numerical solution.

Just as artificial diffusion arises from first-order truncation errors proportional to second space derivatives, dispersion errors typically arise from second-order truncation errors proportional to third space derivatives, or more generally truncation errors proportional to higher space derivatives of odd order. We shall not work out the detailed form of those errors for any particular convective schemes, but their essential features can be inferred from the following general considerations. The first step is to derive the modified differential equation to which the numerical scheme is equivalent by means of a conventional truncation-error analysis. A generic modified equation of this type will be of the form

$$\frac{\partial f}{\partial t} + u\frac{\partial f}{\partial x} = D_a\frac{\partial^2 f}{\partial x^2} + E_a\frac{\partial^3 f}{\partial x^3} + \cdots \tag{7.25}$$

the right member of which represents the truncation errors, since it vanishes for the true differential Eq. (7.1). The dispersive characteristics of the scheme can then be determined by examining the behavior of an individual continuous Fourier mode of the form

$$f = f_0 e^{\iota(\kappa x - \omega t)} \tag{7.26}$$

in the modified equation. Combining Eqs. (7.25) and (7.26), we obtain

$$-\iota\omega + \iota\kappa u = -D_a\kappa^2 - \iota E_a\kappa^3 + \cdots \tag{7.27}$$

so that

$$\omega = \kappa u - \iota D_a\kappa^2 + E_a\kappa^3 + \cdots \tag{7.28}$$

The complex functional relation $\omega(\kappa)$ is referred to as the *disperson relation* for Eq. (7.25). The real and imaginary parts of $\omega = \omega_R + \iota\omega_I$ represent propagation and growth or decay, respectively, as can be seen by rewriting Eq. (7.26) as

$$f = f_0 e^{\omega_I t} e^{\iota(\kappa x - \omega_R t)} \tag{7.29}$$

Just as Eq. (7.2) represents a function propagating with speed u, Eq. (7.29) represents a Fourier mode propagating with a *phase speed* $c(\kappa) = \omega_R/\kappa$, while simultaneously growing or decaying exponentially with a time constant $1/|\omega_I|$. According to Eq. (7.28), the phase speed for Eq. (7.25) is given by

$$c(\kappa) = u + E_a\kappa^2 + \mathcal{O}(\kappa^4) \tag{7.30}$$

which no longer has the constant value u of the exact solution Eq. (7.2), but now depends on κ due to the dispersive truncation errors.

Notice that each spatial derivative in Eq. (7.25) contributes a factor of $\iota\kappa$ in Eq. (7.27), so that higher derivatives of odd order contribute to ω_R and produce dispersion, while those of even order are hyperdiffusional and contribute to ω_I. As long as the difference scheme as a whole is stable, however, one is assured that $\omega_I \leq 0$, so there is usually no need to consider the separate hyperdiffusional terms and their signs individually.

As previously remarked, dispersion errors are often unacceptably large, so the question arises of how best to modify the scheme in order to minimize them. One might at first be tempted to pursue the same type of approach used to deal with artificial diffusion in Sect. 7.3, namely to modify the scheme in such a way as to reduce or remove the lowest-order dispersion errors. Unfortunately, that approach is not very effective for dispersion errors, which tend to be largest for wavelengths on the order of a few Δx. Such wavelengths occur in regions of steep gradients where the solution varies significantly over distances of a few Δx. The Taylor series expansions used to determine the truncation errors will not in general converge rapidly (if at all) in such regions. As a result, the higher-order dispersion errors are likely to be comparable to the lower-order ones, so there is little point in attempting to reduce or remove only the latter. A more fruitful approach is to eschew the truncation-error analysis of dispersion errors in favor of directly examining the numerical errors in the local rate at which f is transported or propagated by the difference scheme in regions of steep gradients. The most transparent approach to the discussion, analysis, and minimization of such errors is via numerical approximation of the associated *fluxes*; i.e., the flow rates per unit area (flow per unit area per unit time) of the conserved quantities such as mass, momentum, and energy. Such fluxes naturally occur and appear when the equations of fluid dynamics are expressed in their fundamental *conservative, conservation,* or *divergence forms*, which directly reflect their origin and significance as local statements of the corresponding conservation laws. *Ceteris paribus*, it is highly desirable for the difference scheme to preserve the conservation properties of the original partial differential equations. The systematic construction of such schemes is discussed in the next section, followed by a brief catalog of some of the methods that have been developed for minimizing dispersion errors by placing various physically motivated constraints on the fluxes.

7.5. Conservative and Finite Volume Schemes

In the absence of source terms, most of the time evolution equations encountered in fluid dynamics can be cast into the conservative or divergence form

$$\frac{\partial \rho_Q}{\partial t} + \nabla \cdot \mathbf{J}_Q = 0 \tag{7.31}$$

where $\rho_Q(\mathbf{x}, t)$ is the local volumetric density of some generic conserved quantity Q, and $\mathbf{J}_Q(\mathbf{x}, t)$ is the corresponding flux vector. The quintessential conserved quantity is the mass M, and the quintessential fluid dynamical conservation equation is the mass continuity equation. The mass density and mass flux are $\rho_M \equiv \rho$ and $\mathbf{J}_M = \rho\mathbf{u}$, respectively, where $\mathbf{u}$ is the fluid velocity vector. The mass continuity equation is then obtained by setting $Q = M$ in Eq. (7.31):

$$\frac{\partial \rho}{\partial t} + \nabla \cdot (\rho\mathbf{u}) = 0 \tag{7.32}$$

The essential ideas of conservative and finite volume schemes can all be elucidated with reference to the continuity equation, to which we may therefore restrict our attention for present purposes. In one-dimensional Cartesian coordinates, Eq. (7.32) reduces to

$$\frac{\partial \rho}{\partial t} + \frac{\partial (\rho u)}{\partial x} = 0 \tag{7.33}$$

When the fluid velocity u is independent of x, Eq. (7.33) further reduces to Eq. (7.1) with f replaced by ρ. In the general case when u depends on x, Eq. (7.33) contains an additional term $\rho(\partial u/\partial x)$, which represents compression or expansion of the fluid. This effect is physically distinct from convection, but in the present context the two terms and effects should not be separated, as this would destroy the conservative form of the equation. The significance of conservation or divergence form is not limited to convection, but in the present context our interest centers on its relevance to minimizing convective dispersion errors.

The form of Eq. (7.31) derives from the fact that conserved quantities such as mass, momentum, and energy cannot be locally created or destroyed, but can merely be transported from one location to another. Difference schemes that preserve those conservation properties are naturally referred to as *conservative schemes*, while those that do not are of course *nonconservative*. As a general rule, nonconservative schemes are undesirable because they artificially create or destroy the conserved quantities, both locally and globally. If the scheme is otherwise accurate, the resulting conservation errors on any single time step will normally be small, but they accumulate from one time step to the next. Unsteady CFD calculations typically proceed for many thousands of time steps, so the accumulation of small errors over many time steps can, and generally does, eventually result in unacceptably large errors in both the total amounts and the spatial distributions of the various conserved quantities within the computational domain.

All rules have exceptions, however, and there are certain situations in which nonconservative schemes are advantageous and are commonly employed, particularly in low-speed flow; i.e., flow at low Mach number, which includes incompressible flow as a special case. In such flows, one typically solves an evolution equation for the internal energy rather than the total energy (which includes the kinetic energy of the fluid). The internal energy equation is simpler in form and generally more convenient to work with. Unlike total energy, internal energy is not a conserved quantity, so its evolution equation is not of the form of Eq. (7.31) and it is not amenable to fully conservative differencing. However, this is a distinction without much of a difference, since the kinetic energy is normally negligible compared to the internal energy at low Mach number. Consequently, the internal energy is nearly conserved and is amenable to the use of nearly conservative difference schemes. Nevertheless, some problems have been reported in which total energy conservation seems essential even at very low Mach number, so this possibility should be kept in mind.

Another exception occurs in low-speed or incompressible flows with large density inhomogeneities. In such flows, the fluid velocity $\mathbf{u}(\mathbf{x}, t)$ tends to be a much smoother function of the position $\mathbf{x}$ than is the momentum density $\rho\mathbf{u}$. Difference approximations to derivatives of $\mathbf{u}$ are then likely to be more accurate than similar approximations to derivatives of $\rho\mathbf{u}$, which mitigates in favor of schemes that solve the momentum equation in nonconservative form. Despite such occasional exceptions, however, conservative schemes are usually preferable and superior to nonconservative schemes. As a result, they are ubiquitous and of primary importance in CFD, and it is essential for even beginning CFD students to acquire a clear understanding of them.

Fortunately, it is straightforward to construct conservative difference schemes that precisely conserve Q to within roundoff errors, both locally and globally. Indeed, local conservation is sufficient, because it implies global conservation. The essence of local conservation is the exchange of equal but opposite amounts of conserved quantities between neighboring volumes or cells, so that whenever an amount of Q enters or leaves a computational cell through any of its boundaries, an equal but opposite amount of Q simultaneously leaves or enters the neighboring cell to which that boundary is common. The manner in which this fundamental physical property follows mathematically from the conservation form of the equations will be worked out in detail below. Of course, conservative schemes remain approximations and have other types of discretization errors, but their conservative structure ensures that those errors cannot manifest themselves as violations of the physical conservation laws.

In sufficiently simple situations, conservative schemes can sometimes be written down by inspection, but the most natural and systematic way of constructing them is by means of the finite-volume method, which automatically preserves the conservation properties of the original partial differential equations. The essence of the finite-volume method is to take the computational cells seriously as well-defined physical subsystems that contain well-defined finite amounts of the various conserved quantities such as mass. Thus, for example, one focuses attention on a particular computational cell and considers the total mass M therein and how it changes with time due to the mass fluxes across its boundaries. In two and three space dimensions, this is done by integrating the conservative evolution equation over the finite volume of a typical computational cell and using the divergence theorem to transform the flux terms into surface integrals over the cell boundaries. This procedure will be described below, but the basic ideas are easier to grasp in one space dimension, where the analogous manipulations involve only ordinary integration and are essentially trivial. We now proceed to work through the steps in this process within the simple context of Eq. (7.33).

Consider the finite-difference cell centered at the point x_{i+}, whose left and right boundaries lie at the points x_i and x_{i+1}, respectively. The index $i+$ also serves as a convenient label that allows us to refer to that particular cell as cell $i+$. The total mass (or more precisely mass per unit area) of cell $i+$ at time t is $M_{i+}(t) = \int_{i+} dx\, \rho(x, t)$, where $\int_{i+} dx$ denotes the integral from x_i to x_{i+1}. It is natural to define the mean density of cell $i+$ by $\rho_{i+} \equiv M_{i+}/\Delta x$. But this obvious definition is more than merely natural–it has the essential consequence that the total mass (again per unit area) of any group of cells, including the entire computational region, is given by the summation of $\rho_{i+}\Delta x$ over those cells.

The time rate of change of M_{i+} can be evaluated by integrating Eq. (7.33) over cell $i+$ to obtain

$$\frac{\partial M_{i+}}{\partial t} + (\rho u)_{i+1} - (\rho u)_i = 0 \tag{7.34}$$

from which it follows that

$$\frac{\partial \rho_{i+}}{\partial t} = -\frac{(\rho u)_{i+1} - (\rho u)_i}{\Delta x} \tag{7.35}$$

A simple fully explicit temporal difference approximation to this equation is given by

$$\frac{\rho_{i+}^{n+1} - \rho_{i+}^n}{\Delta t} = -\frac{(\rho u)_{i+1}^n - (\rho u)_i^n}{\Delta x} \tag{7.36}$$

However, this does not yet represent a unique difference scheme for solving Eq. (7.33), because so far we have defined only the cell densities ρ_{i+}, but not the densities ρ_i at the cell boundaries or mesh points x_i. In contrast, the velocities will be considered to be node-centered variables that are fundamentally located at the mesh points x_i, as discussed in Sect. 2.1, so that the values of u_i can be regarded as known. But even though the scheme is not yet complete, its conservation properties can already be deduced from the structure of Eq. (7.36). This is facilitated by examining Eq. (7.36) together with the corresponding equation for cell $i-$:

$$\frac{\rho_{i-}^{n+1} - \rho_{i-}^n}{\Delta t} = -\frac{(\rho u)_i^n - (\rho u)_{i-1}^n}{\Delta x} \tag{7.37}$$

Now consider the flux $(\rho u)_i^n$, which represents the rate at which mass is flowing across the cell boundary $x = x_i$. The essential point is that this same flux appears in both Eq. (7.36) and Eq. (7.37), but with opposite signs. This implies that regardless of how the densities ρ_i are defined, whatever mass enters or leaves cell $i+$ through the cell boundary x_i simultaneously leaves or enters cell $i-$ through that same common boundary, so that no mass is thereby created or destroyed. This is the essence of conservation. Furthermore, if Eq. (7.36) is multiplied by Δx and summed over all cells then each interior flux $(\rho u)_i^n$ appears twice, again with opposite signs. The interior fluxes consequently cancel in pairs, leaving only the fluxes on the

boundaries of the computational region. The total mass of a closed region
in which the velocities and fluxes vanish on the boundaries is then constant
in time, just as it should be.

The preceding considerations are valid regardless of how ρ_i is defined
on cell boundaries, but the remaining properties of the scheme, in particular
its stability and accuracy, depend critically on that definition. Based upon
the analysis of Sect. 7.1, it is clear that if we were to evaluate ρ_i by
simply setting $\rho_i \equiv \frac{1}{2}(\rho_{i-} + \rho_{i+})$, the resulting FTCS scheme would
be unconditionally unstable. It is further clear from the discussion in
Sects. 7.2 and 7.3 that this defect can be remedied by the use of appropriate
upwinding. In the present context where the fluxes play a central role, it
is both simpler and more fundamental to apply the concept of upwinding
directly to the evaluation of the densities and fluxes on cell boundaries,
rather than to the difference approximations to their spatial derivatives as
was done in Sects. 7.2 and 7.3. Thus when $u_i > 0$, the basic upwind or donor-
cell scheme results from the approximation $\rho_i = \rho_{i-}$, while the conservative
analog of the upwind-weighted difference approximation of Eq. (7.22) is
obtained by setting $\rho_i = \frac{1}{2}(1 + \chi)\rho_{i-} + \frac{1}{2}(1 - \chi)\rho_{i+}$. The corresponding
expressions when $u_i < 0$ are obtained simply by interchanging the indices
$i-$ and $i+$. The basic conservative interpolated upwind scheme analogous to
Eq. (7.24) can then be obtained by determining the local value of χ required
for the artificial diffusivity to vanish, or equivalently by interpolating a
distance $\frac{1}{2}|u|\Delta t$ upstream as described in Sect. 7.3. Either way, one again
obtains the value $\chi = C$. Of course, the resulting scheme remains subject
to the same type of dispersion errors discussed in Sect. 7.4. In and of itself,
the conservative formulation does nothing to ameliorate those errors, but it
does facilitate the development of various methods for doing so by imposing
physically motivated restrictions on the fluxes, as we shall briefly discuss
in the next section.

The concepts and conservation properties discussed above readily
generalize to two and three space dimensions and generalized non-
rectangular meshes. To see how this works, we shall again follow essentially
the same procedure, but starting with the multidimensional Eq. (7.32)
instead of the one-dimensional Eq. (7.33). Let us suppose that the
computational domain in two or three space dimensions has somehow been
subdivided into polygonal or polyhedral computational cells labeled by the
index μ. The faces of those cells will be labeled by the index σ. The total
mass of cell μ at time t is $M_\mu(t) = \int_\mu d\mathbf{x}\, \rho(\mathbf{x}, t)$, where $\int_\mu d\mathbf{x}$ denotes a

spatial area or volume integral over cell μ. The mean density of cell μ is defined by $\rho_\mu = M_\mu/V_\mu$, where $V_\mu \equiv \int_\mu d\mathbf{x}$ is the area or volume of cell μ. Integrating Eq. (7.32) over cell μ and dividing by V_μ, we obtain

$$\frac{\partial \rho_\mu}{\partial t} = -\langle \nabla \cdot (\rho\mathbf{u}) \rangle_\mu \tag{7.38}$$

where

$$\langle g(\mathbf{x}) \rangle_\mu \equiv \frac{1}{V_\mu} \int_\mu d\mathbf{x}\, g(\mathbf{x}) \tag{7.39}$$

is the volume average of an arbitrary function $g(\mathbf{x})$ over cell μ. A fully explicit temporal difference approximation to Eq. (7.38) is then given by

$$\frac{\rho_\mu^{n+1} - \rho_\mu^n}{\Delta t} = -\langle \nabla \cdot (\rho\mathbf{u}) \rangle_\mu^n \tag{7.40}$$

in which $\langle \nabla \cdot (\rho\mathbf{u}) \rangle_\mu^n$ may be regarded as the spatial difference approximation to $\nabla \cdot (\rho\mathbf{u})$ in cell μ at time level n. According to Eq. (7.39), $\langle \nabla \cdot (\rho\mathbf{u}) \rangle_\mu^n$ can be transformed into a surface integral by means of the divergence theorem, with the result

$$\langle \nabla \cdot (\rho\mathbf{u}) \rangle_\mu^n = \frac{1}{V_\mu} \int_\mu dA\, \mathbf{n} \cdot (\rho\mathbf{u})^n = \frac{1}{V_\mu} \sum_{\sigma(\mu)} A_\sigma \mathbf{n}_\sigma \cdot (\rho\mathbf{u})_\sigma^n \tag{7.41}$$

where $\int_\mu dA$ denotes a surface integral over all the faces of cell μ, A_σ is the surface area of cell face σ, $\mathbf{n}_\sigma$ is its outward unit normal vector, $(\rho\mathbf{u})_\sigma \equiv (1/A_\sigma) \int_\sigma dA\, \rho\mathbf{u}$ is the average value of the mass flux $\rho\mathbf{u}$ on cell face σ, and the summation extends over all faces of cell μ. Combining Eqs. (7.40) and (7.41), we obtain

$$\frac{\rho_\mu^{n+1} - \rho_\mu^n}{\Delta t} = -\frac{1}{V_\mu} \sum_{\sigma(\mu)} A_\sigma \mathbf{n}_\sigma \cdot (\rho\mathbf{u})_\sigma^n \tag{7.42}$$

Assuming that the cell-face velocities $\mathbf{u}_\sigma^n$ are again known, all that remains is to define the cell-face densities ρ_σ^n by means of a suitable upwinding scheme. But regardless of how those densities are defined, Eq. (7.42) clearly possesses the same essential conservation properties already discussed in connection with Eq. (7.36). In particular, the flux term $A_\sigma \mathbf{n}_\sigma \cdot (\rho\mathbf{u})_\sigma^n$ also appears in the similar equation for the neighboring cell which shares the cell face σ, but with the opposite sign since from that cell's viewpoint, $\mathbf{n}_\sigma$ points inward. As a result, whatever mass is fluxed into or out of cell μ through a cell face σ is simultaneously fluxed out of or into the neighboring cell on the other side, so that again no mass is created or destroyed. And once again, if Eq. (7.42) is multiplied by V_μ and summed over μ, the interior flux terms cancel in pairs, leaving only the fluxes on the boundaries of the computational region and thereby ensuring the conservation of total mass.

7.6. Flux Limiters and Related Methods

As discussed in Sect. 7.4, the bounded unphysical oscillations that interpolated upwinding schemes typically produce in regions of steep gradients can be attributed to disperson errors. However, they can also be directly analyzed by examining the fluxes that the difference scheme produces in such regions and how they differ from the exact differential fluxes in simple convection problems with known exact solutions. When this is done, the solution irregularities can typically be traced to physically unrealistic values of the fluxes on cell faces, which can be prevented by imposing appropriate constraints, restrictions, or limits on the allowed values of the fluxes. Algorithms of this type are accordingly referred to as *flux limiters*, and they can be designed in such a way as to prevent the occurrence of particular unphysical solution features not present in the exact solution, such as violations of positivity or monotonicity, or the spontaneous creation of new unphysical extrema. Of course, limiting the fluxes is tantamount to modifying the numerical solution, but in a manner that is automatically and inherently conservative. In contrast, attempting to directly detect and remove the unphysical features from the numerical solution itself while preserving local conservation would be a difficult proposition, to say the least.

The development of improved convective difference schemes that incorporate flux-limiting features, including a bewildering variety of flux-limited higher-order upwinding schemes, has become a highly specialized endeavor to which people devote entire careers. We shall make no attempt to discuss even the simplest and most basic aspects of such schemes here. The many and various flux-limiting and related methods include flux-corrected transport (FCT), the Van Leer and superbee limiters, and a wide variety of total variation diminishing (TVD) and essentially non-oscillatory (ENO) schemes. We mention these methods only to call the reader's attention to their existence, and those who are interested will experience no difficulty in rounding up scores if not hundreds of papers on these and similar topics.

Chapter 8

PRESSURE WAVES

It is conventional in fluid dynamics to distinguish between two types of pressure waves: sound waves and shock waves. Strictly speaking there is no sharp distinction between the two, since sound waves can be regarded as infinitely weak shock waves, while conversely shock waves can be regarded as nonlinear sound waves with finite amplitude, but the distinction is nevertheless useful, and in practice there is seldom any ambiguity as to which is which. In contrast to the effects considered in previous chapters, which occur in individual evolution equations, pressure waves arise as a result of coupling between different evolution equations, specifically the momentum equation and the continuity and/or energy equations. Of course, the equations of fluid dynamics are inherently coupled to one another and must be solved together as a system, and the various effects previously discussed in isolation are influenced and modified by that coupling, but they would still occur in its absence, whereas pressure waves owe their very existence to it. Any discussion of numerical methods for pressure waves, even in their simplest and most basic form, consequently requires the consideration and analysis of a system of at least two coupled time evolution equations.

8.1. Sound Waves

The simplest equation system exhibiting sound waves takes the form

$$\frac{\partial \rho}{\partial t} = -\frac{\partial J}{\partial x} \tag{8.1}$$

$$\frac{\partial J}{\partial t} = -c^2 \frac{\partial \rho}{\partial x} \tag{8.2}$$

where $J \equiv \rho u$ has the dual interpretation of mass flux and momentum density. These equations are essentially a partial linearization of the equations for inviscid fluid flow in one space dimension, from which they are easily obtained as follows. First, the continuity and energy equations combine to imply that the flow is isentropic; i.e., $Ds/Dt = 0$, where $s(x,t)$ is the specific entropy (entropy per unit mass) of the fluid and $D/Dt = \partial/\partial t + u\partial/\partial x$ is the one-dimensional convective derivative. It follows that if the initial entropy $s_0(x) \equiv s(x,0)$ is uniform in space, so that s_0 is a constant independent of x, then $s(x,t) = s_0$ for all x and t. The latter isentropic condition can then be used in lieu of the energy equation, which is no longer needed. The remaining evolution equations consist of the one-dimensional continuity and momentum equations. By virtue of the isentropic condition, the pressure gradient in the momentum equation becomes $\partial p/\partial x = c^2\partial\rho/\partial x$, where $c(\rho) = (\partial p/\partial \rho)_s^{1/2}$ is the isentropic sound speed evaluated at $s = s_0$. Finally, one restricts attention to flows in which (a) the fluid velocity u is so small that the convection terms in the momentum equation can be neglected, and (b) density variations are so small that $c(\rho)$ can be treated as a constant c.

Exercise 8.1. Show that the one-dimensional continuity and momentum equations reduce to Eqs. (8.1) and (8.2) under the above conditions.

Note that although the convective momentum flux has been neglected, the convective mass flux has not. This is not inconsistent, however, because the former is quadratic in u while the latter is linear. Equations (8.1) and (8.2) constitute a closed coupled system of linear equations for ρ and J. They can easily be combined to show that both ρ and J satisfy the familiar wave equation

$$\frac{\partial^2 f}{\partial t^2} = c^2 \frac{\partial^2 f}{\partial x^2} \tag{8.3}$$

the general solution of which is

$$f(x,t) = F(x - ct) + G(x + ct) \tag{8.4}$$

where F and G are arbitrary functions which represent waves of arbitrary shape traveling with speed c to the right and left, respectively. This justifies the identification of c with the sound speed.

8.1.1. *Explicit schemes*

As discussed in Sect. 2.1, we shall regard the finite-difference approximations to the densities and velocities as being cell centered and node centered, respectively. It is natural to consider J to be located at the same points as the velocities, so that J too is node centered. The conversion between $J = \rho u$ and u then requires one to approximate the densities at the points x_i, where they are not fundamentally located. However, Eqs. (8.1) and (8.2) are a closed system of equations for ρ and J, so they can be solved as such without explicitly determining the velocities. It is therefore unnecessary for present purposes to define or approximate the densities at the mesh points x_i, so their values at those points can be left ambiguous for now. In more general situations, however, defining ρ_i in terms of $\rho_{i\pm}$ cannot be avoided, and the appropriate definition may differ depending on the context. In particular, when J_i represents the mass flux, as it does in Eq. (8.1), it is often necessary to define ρ_i by a suitable upwinding scheme as discussed in Chapter 7. However, when J_i represents the momentum density, as it does in Eq. (8.2), the centered approximation $\rho_i = \frac{1}{2}(\rho_{i+} + \rho_{i-})$ seems more natural and appropriate. Of course, J simultaneously possesses both interpretations in the differential equations, but in difference schemes it is sometimes essential to distinguish between the two and to represent them by different approximations. This is not improper as long as both approximations are consistent, but it is somewhat regrettable and should be done only when necessary. These issues are best addressed within the context of specific schemes on a case-by-case basis, as will be done in the next chapter.

With the above conventions, an obvious fully explicit FTCS finite-difference approximation to Eqs. (8.1) and (8.2) is

$$\frac{\rho_{i+}^{n+1} - \rho_{i+}^{n}}{\Delta t} = -\frac{J_{i+1}^{n} - J_i^{n}}{\Delta x} \tag{8.5}$$

$$\frac{J_i^{n+1} - J_i^{n}}{\Delta t} = -c^2 \frac{\rho_{i+}^{n} - \rho_{i-}^{n}}{\Delta x} \tag{8.6}$$

This scheme is clearly first-order accurate in time, but by virtue of the spatial centering is second-order accurate in space. Let us proceed to analyze its stability properties. Since the equations are already linear and contain no inhomogeneous terms, the perturbations $\delta\rho_{i+}^{n} = \hat{\rho}\xi^n \exp\{\iota\kappa(i + 1/2)\Delta x\}$ and $\delta J_i^n = \hat{J}\xi^n \exp\{\iota\kappa i\Delta x\}$ satisfy precisely the same equations.

Substituting those perturbations into Eqs. (8.5) and (8.6) and canceling the common factors of $\xi^n \exp\{\iota \kappa i \Delta x\}$, we obtain, after a little algebra,

$$\frac{\xi - 1}{\Delta t} \hat{\rho} = - \left[\frac{2\iota \sin(\kappa \Delta x/2)}{\Delta x} \right] \hat{j} \qquad (8.7)$$

$$\frac{\xi - 1}{\Delta t} \hat{j} = -c^2 \left[\frac{2\iota \sin(\kappa \Delta x/2)}{\Delta x} \right] \hat{\rho} \qquad (8.8)$$

Combining Eqs. (8.7) and (8.8) and solving for ξ, we find

$$\xi(\kappa) = 1 \pm \iota 2C \sin(\kappa \Delta x/2) \qquad (8.9)$$

where $C \equiv c\Delta t/\Delta x$ is the acoustic Courant number. It follows that

$$|\xi(\kappa)|^2 = 1 + 4C^2 \sin^2(\kappa \Delta x/2) \qquad (8.10)$$

and we see that $|\xi(\kappa)| \geq 1$ for all κ and Δt, so that the stability condition of Eq. (4.4) cannot be satisfied for any nonzero value of Δt, however small. The FTCS scheme of Eqs. (8.5) and (8.6) is therefore unconditionally unstable and hence entirely useless, just like the FTCS scheme for convection. This is not surprising when one reflects that both processes are hyperbolic, and indeed are intimately related by virtue of the fact that the wave Eq. (8.3) can be written in the factored form

$$\left(\frac{\partial}{\partial t} + c \frac{\partial}{\partial x} \right) \left(\frac{\partial}{\partial t} - c \frac{\partial}{\partial x} \right) f = 0 \qquad (8.11)$$

which can be interpreted as the product of two oppositely propagating convection equations, as reflected in the exact solutions Eqs. (7.2) and (8.4).

Fortunately, the unconditional instability of Eqs. (8.5) and (8.6) has a simpler and more satisfactory remedy than the use of upwind differencing: it is merely necessary to solve the numerical approximations to Eqs. (8.1) and (8.2) sequentially rather than simultaneously, and evaluate the spatial differences in the latter in terms of the new-time values calculated in the former, or *vice versa*. Both alternatives are equivalent here, but the second turns out to be preferable in more general situations. Thus we retain Eq. (8.6) but replace J_i^n by J_i^{n+1} in Eq. (8.5) (for all i). We thereby obtain the scheme

$$\frac{J_i^{n+1} - J_i^n}{\Delta t} = -c^2 \frac{\rho_{i+}^n - \rho_{i-}^n}{\Delta x} \qquad (8.12)$$

$$\frac{\rho_{i+}^{n+1} - \rho_{i+}^n}{\Delta t} = -\frac{J_{i+1}^{n+1} - J_i^{n+1}}{\Delta x} \qquad (8.13)$$

Equation (8.13) superficially looks implicit, but actually is not since the values of J_i^{n+1} have already been calculated from Eq. (8.12) and hence are available and do not require the solution of a linear algebraic equation system. But the enhanced stability benefits of implicitless nevertheless accrue; Eq. (8.7) of the previous stability analysis is now replaced by

$$\frac{\xi - 1}{\Delta t} \hat{\rho} = -\xi \left[\frac{2\iota \sin(\kappa \Delta x/2)}{\Delta x} \right] \hat{J} \tag{8.14}$$

which combines with Eq. (8.8) to yield

$$\xi^2 + [4C^2 \sin^2(\kappa \Delta x/2) - 2]\xi + 1 = 0 \tag{8.15}$$

Exercise 8.2. Solve Eq. (8.15) for $\xi(\kappa)$ and show that (a) $|\xi(\kappa)| = 1$ for all κ if $C \leq 1$, and that (b) $|\xi(\kappa)| > 1$ for $\kappa = \pi/\Delta x$ if $C > 1$.

The acoustic CFL condition $C = c\Delta t/\Delta x \leq 1$, or $\Delta t \leq \Delta x/c$, is therefore both sufficient and necessary for the scheme of Eqs. (8.12) and (8.13) to be stable. Moreover, when that condition is satisfied the scheme is neutrally stable and manifests no unphysical decay. Owing to its simplicity and favorable stability properties, this scheme is a standard and quite satisfactory explicit scheme for sound waves. It also provides a nice illustration of the benefits of using a staggered mesh, since it involves centered spatial differences that only extend over a single cell and hence are not vulnerable to checkerboarding.

In the present context, there was no need or reason to define either ρ_i or J_i any differently in the difference approximations to Eqs. (8.1) and (8.2), so that J_i was the same quantity in both equations, regardless of whether it was playing the role of mass flux or momentum density. Indeed, since Eqs. (8.1) and (8.2) do not explicitly involve u, it was not even necessary to explicitly define ρ_i at all. But this implies that the schemes analyzed above and their behavior are entirely indifferent to how ρ_i is defined or approximated, and in particular to whether upwinding is used or not. As shown above, the scheme of Eqs. (8.12) and (8.13) is conditionally stable when $c\Delta t/\Delta x \leq 1$ regardless of whether ρ_i is defined in a centered or upwinded manner, so it may as well be approximated by the simple centered expression $\rho_i = \frac{1}{2}(\rho_{i+} + \rho_{i+})$. Thus there is no rationale or motivation for the use of upwinding in the difference approximation to Eq. (8.1), in spite of the fact that Eq. (8.1) is the full continuity equation from which no convection terms have been neglected. One might suspect that this is due to evaluating J_i at the advanced time level in Eq. (8.13), but such is not the case; the scheme obtained by replacing $\rho_{i\pm}^n$ by $\rho_{i\pm}^{n+1}$ in Eq. (8.6) is entirely equivalent and has identical stability properties. Those properties are rather due to the coupling between the equations.

Exercise 8.3. Verify that this coupling results in a cancellation of first-order temporal truncation errors and their associated artificial diffusivities.

These observations have significant implications for more general explicit CFD schemes based on Eqs. (8.12) and (8.13), as they lead to the perhaps surprising conclusion that the convection terms in the continuity equation require no upwinding in properly designed schemes of this type, such as that described in Sect. 9.1 below.

The above procedure of solving coupled equations sequentially rather than simultaneously, and evaluating variables in the current equation in terms of the new-time values obtained from previous equations, is a common practice in CFD, and the effect is generally either benign or beneficial, as it was above. This would intuitively be expected on the basis that composite schemes of this type inherently make use of all the currently available information at the advanced time level, which would be expected to produce some of the same general benefits that implicitness does, as discussed in Sect. 5.1.4.

8.1.2. *Implicit schemes*

The equations of fluid dynamics contain two basic propagation speeds: the fluid speed $|u|$ and the sound speed c. As long as those two speeds are of the same order of magnitude, their respective explicit stability limits are comparable and there is no motivation to consider the use of implicit schemes for treating the sound waves. However, *low-speed* fluid flows for which $|u| \ll c$ are very common. The associated dimensionless parameter is the *Mach number* Ma $= |u|/c$, so that "low-speed" is synonymous with "low Mach number." As a general rule, Mach numbers on the order of 0.1 or less are considered to be low, in which case the flow may be regarded as essentially incompressible. A low Mach number implies a wide disparity or separation between the convective and acoustic time scales and their associated explicit stability limits. As a result, the time step in explicit calculations at low Mach number is controlled by the acoustic CFL condition $\Delta t \leq \Delta x/c$. This is much more restrictive than the convective CFL condition $\Delta t \leq \Delta x/|u|$, and consequently requires the use of much smaller time steps than could otherwise be taken.

One is therefore led to contemplate the use of implicit schemes for sound waves in hopes of thereby avoiding the acoustic CFL stability condition. Of course, this would not be sensible if the latter condition were also an accuracy condition, but this is often not the case at low Mach number, where the amplitudes of the pressure and velocity variations associated with sound waves are normally very small. The sound waves are then basically tiny ripples superimposed on the underlying essentially incompressible flow, upon which they have no significant effect. When the details of those ripples are not themselves of interest (e.g., when they are produced by the incoherent mutterings of those who disagree with us), there is no need to compute them accurately, and one is more than willing to sacrifice their accuracy in exchange for a larger time step by using implicit schemes. This situation is the epitome of *stiffness*, which generally connotes a wide separation between the relatively long physical time scales of interest and a much shorter characteristic time associated with an insignificant or irrelevant phenomenon of small amplitude.

The above considerations constitute the rationale for using implicit schemes to simulate sound waves. We shall restrict our attention to the obvious fully implicit scheme obtained by replacing $\rho_{i\pm}^n$ by $\rho_{i\pm}^{n+1}$ in Eq. (8.12). We thereby obtain the scheme

$$\frac{\rho_{i+}^{n+1} - \rho_{i+}^n}{\Delta t} = -\frac{J_{i+1}^{n+1} - J_i^{n+1}}{\Delta x} \tag{8.16}$$

$$\frac{J_i^{n+1} - J_i^n}{\Delta t} = -c^2 \frac{\rho_{i+}^{n+1} - \rho_{i-}^{n+1}}{\Delta x} \tag{8.17}$$

The stability analysis proceeds in the usual way, and differs from the preceding one only in that Eq. (8.8) is now replaced by

$$\frac{\xi - 1}{\Delta t} \hat{J} = -\xi c^2 \left[\frac{2\iota \sin(\kappa \Delta x/2)}{\Delta x} \right] \hat{\rho} \tag{8.18}$$

which combines with Eq. (8.14) to yield

$$[1 + 4C^2 \sin^2(\kappa \Delta x/2)]\xi^2 - 2\xi + 1 = 0 \tag{8.19}$$

Solving for ξ, we obtain

$$\xi = \frac{1 \pm \iota 2C \sin(\kappa \Delta x/2)}{1 + 4C^2 \sin^2(\kappa \Delta x/2)} \tag{8.20}$$

from which it readily follows that

$$|\xi|^2 = \frac{1}{1 + 4C^2 \sin^2(\kappa\Delta x/2)} \tag{8.21}$$

so that $|\xi| \leq 1$ for all values of κ and Δt. As anticipated, the scheme of Eqs. (8.16) and (8.17) is therefore unconditionally stable and is no longer subject to the acoustic CFL stability condition.

8.2. Shock Waves

Shock waves are an important feature of many transsonic and supersonic flows. By their very nature, they are characterized by steep gradients in the dependent variables, indeed infinitely steep in the inviscid limit, in which the solutions become discontinuous. Shock waves and their effects are consequently difficult to accurately represent in finite-difference schemes, with their inherently finite resolution. Numerical simulations of fluid flows involving shock waves consequently require the use of special techniques.

In the continuum equations of fluid dynamics, shock waves can be described in two different ways. In the inviscid equations (i.e., when the viscosities and thermal conductivity are taken to be zero), shock waves are propagating discontinuities in the dependent variables, the magnitudes of which are determined by the Rankine–Hugoniot jump conditions. Those conditions connect the piecewise continuous solutions on either side of the discontinuity, and hence essentially serve as internal boundary conditions. They also determine the speed of the shock wave relative to the fluid, and thereby its location. The jump conditions are direct consequences of the conservation properties of the governing partial differential equations. As a result, the shock speed and jump conditions, and hence the entire solution, depend critically on those underlying conservation laws.

The second continuum description of shock waves employs the equations of viscous fluid flow (the Navier–Stokes equations), in which the momentum and energy equations include viscous stresses, as well as heat fluxes resulting from thermal conduction. Viscous stresses and heat conduction are both diffusive effects, and they have the effect of smoothing out discontinuous shock waves into thin but very steep transition layers of finite width, across which the dependent variables vary rapidly but continuously. This width is typically only several molecular mean free paths, which is much smaller than the cell size in practical finite-difference calculations. This implies that shock waves cannot normally be directly resolved as smooth transitions in such calculations. Indeed, shock waves are generally so thin that their internal structure cannot even be accurately described by the usual partial differential equations of fluid dynamics, which themselves become inaccurate at molecular length scales. Fortunately, those inaccuracies are ordinarily harmless, since the continuous differential equations do at least predict qualitatively reasonable shock profiles with about the right thickness. The detailed internal structure of a shock wave is not normally of interest as long as its speed and the jumps across it are accurate. The latter features are determined by the conservation laws, so they are accurately captured by the continuum conservation equations.

There are two basic types of numerical methods for the treatment of shock waves, namely *shock fitting* and *shock capturing* methods. These roughly correspond to the two different continuum descriptions discussed above. Shock fitting methods treat shock waves as true discontinuities and use the jump conditions to determine their propagation velocities and the matching conditions that must be imposed to connect the solutions on either side. Such methods are complex and challenging, especially in two and three space dimensions, for several reasons. First, they obviously require special interpolation logic to deal with the fact that shock waves do not normally coincide with cell boundaries, as well as the use of special one-sided difference approximations on either side of the shock. A second and more serious complication is that they require *a priori* knowledge of the initial location of the shock wave, which in many cases is unavailable because shocks can spontaneously evolve from initially smooth solutions due to the self-steepening property of the nonlinear convection terms in the momentum equation. General-purpose shock-fitting methods therefore require special logic to detect spontaneous shock formation, which presents a difficult problem. The treatment of shock reflections, intersections, branching, and merging present further difficult problems. The primary compensating advantage of shock-fitting methods is that they represent shock waves as sharp discontinuities. In contrast, shock-capturing methods are much simpler and treat all of the above complications automatically, but at the cost of artificially spreading the shock waves out over a few finite-difference cells.

Shock-capturing methods treat shocks as continuous but thin transition layers produced by viscous effects, so that the shock speed and jump conditions need not be explicitly imposed but emerge automatically from the calculation. As mentioned above, however, physical values of the molecular viscosity and thermal conductivity typically produce shock widths that are far too thin to resolve in a practical finite-difference mesh. Merely solving the viscous fluid dynamics equations by standard finite-difference methods is therefore inadequate to capture shock waves in a physically realistic manner. In order for a shock wave to be represented as a smooth transition layer in a finite-difference mesh, it must be artificially thickened to the point where it can be at least coarsely resolved in the mesh. This requires an artificial thickness on the order of a few Δx. Moreover, this must be accomplished in such a way that neither the shock speed nor the jump conditions are altered by the artificial thickening. As discussed above, the latter features are direct consequences of the conservation properties of the partial differential equations. This implies that *shock-capturing methods require the use of conservative difference schemes.* This requirement is absolutely essential to obtain correct shock speeds and jump conditions, and it cannot be emphasized too strongly.

The remaining requirement that the shock must be artificially thickened may then be satisfied by exploiting the fact that the continuum equations predict a shock thickness proportional to a linear combination of the shear and bulk viscosities and the thermal conductivity. The required thickening can then be accomplished either by the use of artificially large values of the latter parameters, or by the use of difference schemes with viscous or dissipative truncation errors. There is no need for an artificial thermal conductivity, since viscosity alone is sufficient to produce a continuous transition layer. Indeed, there is no need for an artificial shear viscosity either; an artificial bulk viscosity (or equivalent truncation errors) suffices, which reflects the fact that shock waves are an essentially compressive phenomenon. An artificial bulk viscosity ζ produces an artificial pressure of the form $q = -\zeta \nabla \cdot \mathbf{u}$. This was precisely the form of the earliest shock-capturing method due to von Neumann and Richtmyer (VNR). Since ζ is introduced for purely numerical purposes, it need not be even approximately constant but can be chosen to have any convenient form that produces the desired behavior. Expansion shocks do not exist, so it is customary to set $\zeta = 0$ when $\nabla \cdot \mathbf{u} > 0$. When $\nabla \cdot \mathbf{u} < 0$, it turns out that making ζ proportional to $|\nabla \cdot \mathbf{u}|$ produces an artificial shock thickness nearly independent of shock strength, and the coefficient of proportionality can then be set to a value that produces the desired thickness of a few Δx. The resulting artificial pressure q is then quadratic in $\nabla \cdot \mathbf{u}$, whereas its physical analog is linear. As a result, q is negligibly small in smooth regions of the flow where velocities are smooth and their gradients are small, but in effect automatically turns itself on in shock waves, where the velocity jump produces much larger velocity gradients.

The VNR method is now regarded as obsolete in some quarters, but it still remains useful and has the timeless advantages of great simplicity and a clear physical interpretation. It was followed by the development of other shock-capturing methods based on conservative difference schemes, in which a similar shock-smearing effect is achieved by dissipative truncation errors in the scheme itself rather than the explicit introduction of an artificial viscosity. Noteworthy schemes of this type include those of Lax & Wendroff, MacCormack, and Godunov, variants of which are still widely used. More recent shock-capturing schemes tend to be based on higher-order upwinding combined with flux limiters. The convection terms are essential in shock waves, and together with the presence of steep gradients, this implies that the various flux-limiting and related methods discussed in Sect. 7.6 are directly and equally relevant to shock capturing. Indeed, many such methods were motivated by that application, where they are sometimes needed to prevent artificial oscillations or "ringing" behind the shock wave due to dispersion errors.

The artificial viscosities introduced in shock-capturing methods, either directly as in the VNR method or indirectly by means of dissipative truncation errors, are typically many orders of magnitude larger than molecular viscosities, so such methods are entirely indifferent to whether the original partial differential equations are inviscid or not. Shock-capturing methods are consequently equally applicable in both inviscid and viscous fluid flows, and will produce essentially the same effect and results in both cases. In either case, their purpose and function is simply to artificially thicken shock waves to the degree necessary for them to be resolvable as smooth transitions in the computing mesh. Once this has been done, it is irrelevant whether the shocks were represented as discontinuities or as continuous but unresolvable thin transition layers in the original differential equations.

Chapter 9

COMBINING THE ELEMENTS

It has not been our purpose to propose, develop, or recommend specific unified composite numerical schemes suitable for computing approximate numerical solutions to particular problems described by particular systems of fluid dynamical equations. Rather, our aim has been to provide the reader with some of the essential fundamental background and insights that are prerequisites for the intelligent development, modification, and use of such schemes. Nevertheless, it seems desirable to provide some simple illustrations of how the various ingredients we have discussed can be combined to obtain a complete composite numerical method. We shall do this by showing how to construct some simple basic numerical schemes for compressible flow, incompressible flow, and low-speed flow; i.e., compressible flow at very low Mach number, where the wide separation between the convective and acoustic time scales makes explicit schemes inefficient and impractical (as well as vulnerable to roundoff errors). The schemes discussed below are not state-of-the-art by any means, and yet they can actually be used to generate meaningful solutions to real fluid dynamics problems. The development of CFD schemes, like most human endeavors, is evolutionary, and the most effective known way to proceed is to start with something simple that works but is insufficiently robust, accurate, and/or efficient, build on and modify it, and incrementally refine and improve it by sequentially removing its most conspicuous deficiencies when they reveal themselves, as they inevitably will do. The schemes given below provide eminently reasonable starting points for such further refinements.

9.1. Inviscid Compressible Flow

In the absence of gravity or other external forces, the governing equations for inviscid compressible flow of a pure fluid in one space dimension can be written in conservation form as

$$\frac{\partial \rho}{\partial t} + \frac{\partial (\rho u)}{\partial x} = 0 \tag{9.1}$$

$$\frac{\partial (\rho u)}{\partial t} + \frac{\partial (\rho u^2)}{\partial x} = -\frac{\partial p}{\partial x} \tag{9.2}$$

$$\frac{\partial (\rho E)}{\partial t} + \frac{\partial (\rho E u)}{\partial x} = -\frac{\partial (p u)}{\partial x} \tag{9.3}$$

where ρ is the density, u is the velocity, $E = \frac{1}{2}u^2 + e$ is the total (kinetic plus internal) energy per unit mass, e is the specific internal energy, and p is the pressure, which is given by an equation of state of the form $p = \phi(\rho, e)$. The right member of Eq. (9.3) represents the rate of change of the total energy density due to the work done by pressure forces.

For historical reasons, Eqs. (9.1)–(9.3) are referred to as the *Eulerian* equations of motion. This terminology connotes that they are written in terms of a fixed spatial coordinate system or laboratory frame of reference. The Eulerian equations can be transformed into equivalent *Lagrangian* equations of motion by means of a time-dependent transformation to spatial coordinates that move with the fluid velocity. More general equations can be obtained by the use of an arbitrary time-dependent coordinate transformation. The numerical solution of such transformed equations by finite-difference methods gives rise to Lagrangian schemes, in which the finite-difference mesh moves with the fluid, and more general schemes in which the mesh moves in an arbitrary prescribed manner. Schemes of the latter type are often referred to as arbitrary Lagrangian-Eulerian or ALE schemes. The prescription for moving the mesh can either be explicitly specified, or it can be *adaptive*; i.e., it can be defined in terms of the evolving solution features, typically to obtain improved accuracy by concentrating mesh points and resolution in regions of steep gradients. Schemes based on moving meshes are widely used for particular applications, but are beyond the scope of this book.

As before, we shall suppose that the densities, as well as the other thermodynamic variables e and p, are cell centered, while velocities are node centered. With these conventions, a simple conservative explicit difference scheme for solving Eqs. (9.1)–(9.3) can be constructed by combining appropriate ingredients discussed in previous chapters. The heart of the scheme is a natural generalization of the basic scheme of Eqs. (8.12) and (8.13) for treating sound waves:

$$\frac{(\rho u)_i^{n+1} - (\rho u)_i^n}{\Delta t} = -\frac{(\rho u^2)_{i+}^n - (\rho u^2)_{i-}^n}{\Delta x} - \frac{p_{i+}^n - p_{i-}^n}{\Delta x} \tag{9.4}$$

$$\frac{\rho_{i+}^{n+1} - \rho_{i+}^n}{\Delta t} = -\frac{(\rho u)_{i+1}^{n+1} - (\rho u)_i^{n+1}}{\Delta x} \tag{9.5}$$

Equation (9.4) determines the advanced-time values of the momentum density $(\rho u)_i^{n+1}$ at the mesh points x_i, whereupon Eq. (9.5) then determines the new-time densities ρ_{i+}^{n+1} at the cell centers x_{i+}. However, Eq. (9.4) requires evaluation of the convective momentum fluxes $(\rho u^2)_{i\pm}^n$ at the cell centers $x_{i\pm}$, where the velocities $u_{i\pm}$ are not fundamentally located and hence must be approximated. Those fluxes are evaluated at the old time level n, so they constitute an explicit difference approximation to convective effects, and the analysis of Chapter 7 then suggests that some degree of upwinding will be required for stability. An appropriate upwinding formulation in this context requires a little thoughtful contemplation, because the two factors of u in the convective momentum flux ρu^2 actually play two different roles physically. The conceptual significance of the flux ρu^2 is easier to see if it is thought of and written as $(\rho u)u$, in which the momentum density (i.e., momentum per unit volume) ρu is the density of the conserved quantity that is being convected, while the other factor of u is simply the fluid velocity that is doing the convecting. This interpretation makes it clear that only the factor $(\rho u)_{i\pm}^n$ requires upwinding for stability, while the remaining factor of $u_{i\pm}^n$ might as well be evaluated by means of the centered approximation $u_{i\pm}^n = \frac{1}{2}(u_i^n + u_{i\pm1}^n)$. We therefore evaluate the convective momentum fluxes in Eq. (9.4) as

$$(\rho u^2)_{i\pm}^n = \frac{1}{2}(u_i^n + u_{i\pm1}^n)(\rho u)_{i\pm}^n \tag{9.6}$$

in which the momentum densities $(\rho u)_{i\pm}^n$ at cell centers are evaluated in terms of the known nodal values $(\rho u)_i^n$ by some appropriate upwinding scheme. Once such a scheme has been selected, all quantities appearing in Eqs. (9.4) and (9.5) are well defined, and those equations then determine $(\rho u)_i^{n+1}$ and ρ_{i+}^{n+1}. In accordance with the discussion of Sect. 8.1.1, we further define ρ_i at the mesh points x_i by the simple centered approximation $\rho_i = \frac{1}{2}(\rho_{i+} + \rho_{i-})$, and this in turn determines $u_i^{n+1} = (\rho u)_i^{n+1}/\rho_i^{n+1}$.

All that remains is to construct a suitable explicit conservative difference approximation to the energy equation Eq. (9.3), from which e_{i+}^{n+1} can be determined. The state equation then determines $p_{i+}^{n+1} = \phi(\rho_{i+}^{n+1}, e_{i+}^{n+1})$, thereby completing the time advancement from level n to level $n+1$. In order to proceed, we must first decide whether to approximate Eq. (9.3) at the mesh points x_i or the cell centers $x_{i\pm}$. Either alternative entails unpalatable compromises (which alas abound in CFD), since the total energy density ρE is a hybrid variable comprised of the cell-centered thermodynamic variables ρ and e and the node-centered velocity u, and hence is neither fish nor fowl. In the present context, however, the essential function of the energy equation Eq. (9.3) is to determine the internal energies e_{i+}^{n+1} at cell centers, which is greatly facilitated by approximating Eq. (9.3) at the same points. With this in mind, and remembering that u_i^{n+1} has already been calculated as described above, the obvious basic scheme of the general type discussed in Sect. 7.5 is

$$\frac{(\rho E)_{i+}^{n+1} - (\rho E)_{i+}^n}{\Delta t} = -\frac{(\rho E)_{i+1}^n u_{i+1}^{n+1} - (\rho E)_i^n u_i^{n+1}}{\Delta x}$$
$$- \frac{p_{i+1}^n u_{i+1}^{n+1} - p_i^n u_i^{n+1}}{\Delta x} \qquad (9.7)$$

Equation (9.7) involves several variables at points where they are not fundamentally located and must consequently be approximated. Apart from the work term (which requires separate consideration and will be discussed below), Eq. (9.7) is essentially convective in nature, and it involves ρE at both nodes and cell centers, both of which must be approximated. It determines the advanced-time values of ρE at cell centers, so it is natural to regard ρE as being a cell-centered composite variable. We shall therefore proceed by first approximating ρE at cell centers, where it is given by

$$(\rho E)_{i\pm} = \frac{1}{2}(\rho u^2)_{i\pm} + (\rho e)_{i\pm} \qquad (9.8)$$

Since both ρ and e are cell-centered quantities, $(\rho e)_{i\pm} = \rho_{i\pm} e_{i\pm}$ is already well defined and requires no further approximation or discussion. However, the kinetic energy density $\frac{1}{2}\rho u^2$ at cell centers involves velocities at cell centers, where they are not fundamentally located and must consequently be approximated. This can be done in various ways, of which the obvious simple choice is again $u_{i\pm} = \frac{1}{2}(u_i + u_{i\pm 1})$, so that $\frac{1}{2}(\rho u^2)_{i\pm} = \frac{1}{2}\rho_{i\pm} u_{i\pm}^2$. With this choice, all quantities appearing in Eq. (9.8) have now been defined, which completes the approximation of ρE at cell centers.

It is also necessary to approximate ρE at the nodes x_i in terms of its values at cell centers, so that the convective terms in Eq. (9.7) can be evaluated. The analysis of Chapter 7 again suggests that some degree of upwinding will be necessary for stability. We shall therefore simply suppose that some appropriate upwinding scheme, of the type discussed in Chapter 7, has been used to define $(\rho E)_i^n$ in terms of the known quantities $(\rho E)_{i\pm}^n$ given by Eq. (9.8).

Finally, we must approximate the nodal pressures p_i appearing in the work term in terms of the cell-centered pressures $p_{i\pm}$. This requires us to sail into uncharted waters, since the work term is not among the basic ingredients considered in preceding chapters. At first glance, there would seem to be no obvious reason why p_i therein should not simply be approximated in a centered or symmetrical manner by setting $p_i^n = \frac{1}{2}(p_{i+}^n + p_{i-}^n)$. The latter definition has indeed been widely employed. However, L. D. Cloutman (private communication) has argued that this centered approximation to the nodal pressures introduces a weak numerical instability, and that the simplest and most natural way to restore stability is to evaluate p_i^n by the same upwinding scheme used to define $(\rho E)_i^n$. The fact that this instability is rarely observed in practice does not of course disprove its existence, since such instabilities are often masked or overshadowed by a surplus of dissipation in other parts of the numerical scheme. This particular instability, and the need for upwinding to remove it, are suggested by the observation that although the work term is not physically convective in nature, it nevertheless shares the same mathematical form and character as the convection term, and hence would be expected to have similar stability properties. The resemblance between the two is further enhanced by the fact that the variables p and ρe are very closely related, and indeed are simply proportional to each other in ideal gases with constant specific heats, in which case pu looks formally like an additional convective flux of internal energy. Cloutman reports that these intuitive plausibility considerations are supported by both truncation-error and Fourier stability analyses, as well as computational evidence. We shall therefore follow his lead and suppose that p_i^n has been defined by the same upwinding scheme as $(\rho E)_i^n$.

All quantities appearing in Eq. (9.7) have now been defined, so that equation can now be explicitly solved for $(\rho E)_{i+}^{n+1}$. This provides all the remaining information required to complete the time advancement from level n to level $n + 1$. The overall structure of a computational cycle is therefore as follows:

- Solve Eq. (9.4) to obtain $(\rho u)_i^{n+1}$.
- Solve Eq. (9.5) to obtain ρ_{i+}^{n+1}.
- Compute $\rho_i^{n+1} = \frac{1}{2}(\rho_{i+}^{n+1} + \rho_{i-}^{n+1})$.
- Compute $u_i^{n+1} = (\rho u)_i^{n+1}/\rho_i^{n+1}$ and $u_{i\pm}^{n+1} = \frac{1}{2}(u_i^{n+1} + u_{i\pm1}^{n+1})$.
- Solve Eq. (9.7) to obtain $(\rho E)_{i+}^{n+1}$.
- Compute $(\rho e)_{i+}^{n+1} = (\rho E)_{i+}^{n+1} - \frac{1}{2}(\rho u^2)_{i+}^{n+1}$.
- Compute $e_{i+}^{n+1} = (\rho e)_{i+}^{n+1}/\rho_{i+}^{n+1}$.
- Compute $p_{i+}^{n+1} = \phi(\rho_{i+}^{n+1}, e_{i+}^{n+1})$.

As discussed in Sect. 4.3, we would expect the overall stability condition for the scheme of Eqs. (9.4)–(9.8) to be reasonably well approximated by combining Eq. (4.12) with the explicit convective and acoustic stability limits derived in Sects. 7.2, 7.3, and 8.1.1. We thereby obtain

$$\Delta t < \left(\frac{|u|}{\Delta x} + \frac{c}{\Delta x}\right)^{-1} = \frac{\Delta x}{|u| + c} \tag{9.9}$$

in which a safety factor somewhat less than unity should be inserted as discussed in Chapter 4.

The scheme of Eqs. (9.4)–(9.8) has been constructed in one space dimension for simplicity, but its generalization to two or three space dimensions is straightforward. In fact, schemes of this type are actually of limited interest in one space dimension, where competing Lagrangian schemes are generally preferable. The reason is that Lagrangian schemes eliminate the Eulerian convection terms and their associated artificial diffusion. In two or three space dimensions, this desirable property is unfortunately counterbalanced by the equally serious disadvantage of mesh tangling, which is topologically impossible in one dimension. Mesh tangling can be ameliorated by the use of ALE schemes, which are considerably more complicated and are beyond the scope of this book. The net result is that the much simpler Eulerian schemes are more common in two and three dimensions, while Lagrangian schemes are comparably simple only in one dimension.

9.2. Viscous Low-Speed Flow

In the absence of external forces or heat sources, the governing equations for viscous compressible flow in any number of spatial dimensions can be written in vector form as

$$\frac{\partial \rho}{\partial t} + \nabla \cdot (\rho \mathbf{u}) = 0 \tag{9.10}$$

$$\frac{\partial (\rho \mathbf{u})}{\partial t} + \nabla \cdot (\rho \mathbf{u} \mathbf{u}) = -\nabla p + \nabla \cdot \boldsymbol{\tau} \tag{9.11}$$

$$\frac{\partial (\rho e)}{\partial t} + \nabla \cdot (\rho e \mathbf{u}) = -p \nabla \cdot \mathbf{u} + \boldsymbol{\tau} : \nabla \mathbf{u} - \nabla \cdot \mathbf{J}_q \tag{9.12}$$

where $\boldsymbol{\tau}$ is the viscous stress tensor, which has the usual Newtonian form proportional to products of viscosities and velocity gradients, $\mathbf{J}_q$ is the energy flux due to heat conduction, which has the usual Fourier form proportional to the product of the thermal conductivity and the temperature gradient ∇T, and the other symbols have their usual previously defined meanings. In order to comprise a closed system, these equations require state relations of the form $p(\rho, e)$ and $T(\rho, e)$, as well as expressions for the viscosities and thermal conductivity as functions of (ρ, e) or (ρ, T). As discussed in Sect. 7.5, there is normally no significant disadvantage to the use of the internal energy equation rather than the total energy equation in the present context.

As was already discussed in Sect. 8.1.2, the numerical solution of Eqs. (9.10)–(9.12) at low Mach number is problematical due to the wide separation between the convective and acoustic time scales. This scale separation is reflected in the fact that the explicit CFL sound-speed stability limit $c \Delta t / \Delta x < 1$ is much more restrictive than the typical explicit convective stability limit $|\mathbf{u}| \Delta t / \Delta x < 1$, and this disparity makes the use of explicit numerical schemes inefficient at low Mach number. The inefficiency can be ameliorated by the use of implicit schemes that allow the use of time steps that violate the acoustic CFL condition. The sound waves themselves are then no longer accurately computed, but this is usually immaterial since they are typically of small amplitude and no interest.

The crucial aspect of such implicit schemes for low-speed flow is the detailed structure of the time differencing. The spatial differencing is of secondary importance as long as it is done in a reasonable manner, which as we have seen may require upwinding in the convection terms. We shall accordingly suppress most of the details of the spatial differencing in order to focus more clearly on the time differencing. To be somewhat specific, however, we shall suppose that the spatial difference approximations are constructed by the finite-volume method discussed in Sect. 7.5. The finite-difference cells and their faces will again be labeled by the indices

μ and σ, respectively, and the thermodynamic variables ρ, e, p, and T are located at cell centers while the velocities $\mathbf{u}$ are located on cell faces. (In some schemes, especially in rectangular meshes, only the normal velocity components are located on the cell faces, so that different velocity components are located on different faces.) The finite-volume method then requires the construction of a dual mesh comprised of *momentum cells* whose centers lie on the faces of the original cells. (Indeed, multidimensional CFD schemes often simultaneously employ more than one type of dual mesh and momentum cells.) The original and momentum cells together then constitute a staggered mesh. Again, however, we emphasize that other spatial difference approximations can equally well be used in conjunction with the basic time differencing discussed below.

The simplest implicit schemes for low-speed flow are partially implicit schemes that deliberately introduce only the minimum implicitness required to remove the stability condition $c\Delta t/\Delta x < 1$. The stability properties of the various schemes considered in Chapter 8, in particular that of Eqs. (8.16) and (8.17), suggest that it should be sufficient for this purpose to appproximate the continuity equation and the pressure gradient terms in the momentum equation implicitly, and all remaining terms explicitly. We are thereby led to consider the following temporal difference approximation to Eqs. (9.10) and (9.11):

$$\frac{\rho_\mu^{n+1} - \rho_\mu^n}{\Delta t} + \langle \nabla \cdot (\rho\mathbf{u}) \rangle_\mu^{n+1} = 0 \tag{9.13}$$

$$\frac{(\rho\mathbf{u})_\sigma^{n+1} - (\rho\mathbf{u})_\sigma^n}{\Delta t} + \langle \nabla \cdot (\rho\mathbf{u}\mathbf{u}) \rangle_\sigma^n = -\langle \nabla p \rangle_\sigma^{n+1} + \langle \nabla \cdot \boldsymbol{\tau} \rangle_\sigma^n \tag{9.14}$$

in which the spatial differencing denoted by $\langle \cdots \rangle$ is done by the finite-volume method, as discussed in Sect. 7.5. Based on the previous discussion in Sects. 8.1 and 9.1, we would expect that upwinding will be required in $\langle \nabla \cdot (\rho\mathbf{u}\mathbf{u}) \rangle_\sigma^n$, but not in $\langle \nabla \cdot (\rho\mathbf{u}) \rangle_\mu^{n+1}$. In contrast to the corresponding explicit scheme of Eqs. (9.4) and (9.5), however, Eqs. (9.13) and (9.14) do not constitute a closed system for ρ_μ^{n+1} and $\mathbf{u}_\sigma^{n+1}$. The reason is that $p_\mu^{n+1} = \phi(\rho_\mu^{n+1}, e_\mu^{n+1})$ involves e_μ^{n+1}, which has not yet been computed. This deficiency could be remedied by solving Eqs. (9.13) and (9.14) simultaneously with the finite-difference approximation to Eq. (9.12), but this would greatly complicate the scheme. Indeed, if this were done one might as well go all the way to a fully implicit scheme, since once the scheme requires one to solve for ρ_μ^{n+1}, $\mathbf{u}_\sigma^{n+1}$, and e_μ^{n+1} simultaneously, all the terms in the equations could be evaluated at the advanced-time level without introducing any further additional unknowns. In the interests of simplicity, one is therefore led to approximate p_μ^{n+1} in Eq. (9.14) in terms of ρ_μ^{n+1} as best one can without knowledge of e_μ^{n+1}. The approximation that immediately suggests itself is

$$p_\mu^{n+1} \approx p_\mu^n + \left(\frac{\partial p}{\partial \rho}\right)_\mu^n (\rho_\mu^{n+1} - \rho_\mu^n) \tag{9.15}$$

where the partial derivative $(\partial p/\partial \rho)$ is evaluated at constant θ, and θ is some other thermodynamic variable that we are free to select on physical grounds. The obvious choices for θ are s, T, and e, all of which have been proposed at one time or another. Physically, the best choice for θ is whatever thermodynamic variable will change the least due to the changes in ρ and e occurring on the current time step, but this is not usually known *a priori*. If the viscosity and thermal conductivity are so small that the flow is nearly isentropic, then $\theta = s$ is appropriate, whereas if the thermal conductivity is so large that the flow is nearly isothermal then $\theta = T$ may be preferable. It is clear that no single choice of θ will be well suited for all problems, so the choice should be made based on the physics of the particular flow. The choice of θ is a matter of accuracy rather than consistency, since Eq. (9.15) is first-order accurate in time regardless of how θ is defined. However, a poor choice of θ can also result in numerical instability.

Thus it will henceforth be understood that p_μ^{n+1} in Eq. (9.14) is to be approximated by means of Eq. (9.15) with some appropriate choice of θ. Equations (9.13) and (9.14) then constitute a closed coupled linear system of equations for ρ_μ^{n+1} and $(\rho \mathbf{u})_\sigma^{n+1}$. As discussed in Sects. 3.3 and 3.4, those equations are normally solved iteratively. Once this has been done, the velocities themselves are given by $\mathbf{u}_\sigma^{n+1} = (\rho \mathbf{u})_\sigma^{n+1}/\rho_\sigma^{n+1}$, where the face-centered densities ρ_σ^{n+1} must again be approximated in terms of the cell-centered densities ρ_μ^{n+1}, just as was done in the previous section.

All that remains is to adopt a suitable difference approximation for the energy equation Eq. (9.12), of which perhaps the simplest choice is

$$\frac{(\rho e)_\mu^{n+1} - (\rho e)_\mu^n}{\Delta t} + \langle \nabla \cdot (\rho e \mathbf{u}) \rangle_\mu^n =$$
$$- \langle p\nabla \cdot \mathbf{u} \rangle_\mu^n + \langle \boldsymbol{\tau} : \nabla \mathbf{u} \rangle_\mu^n - \langle \nabla \cdot \mathbf{J}_q \rangle_\mu^n \tag{9.16}$$

This equation explicitly determines $(\rho e)_\mu^{n+1}$, and hence $e_\mu^{n+1} = (\rho e)_\mu^{n+1}/\rho_\mu^{n+1}$, without further ado. If desired, the velocities in Eq. (9.16) could equally well be evaluated at time level $n + 1$ rather than n, since the advanced-time velocities have already been calculated, but this is not critical. Either way, the convective terms in Eq. (9.16) are again treated explicitly, so they will again require some degree of upwinding for stability. In contrast to the scheme of the previous section, the $p\nabla \cdot \mathbf{u}$ work term in the present scheme involves only the cell-centered pressures p_μ, and hence requires no upwinding. This feature is a direct consequence of solving the evolution equation for the internal energy rather than the total energy. Finally, the thermodynamic equation of state determines the advanced-time pressures in terms of the advanced-time densities and internal energies, thereby completing the time advancement from level n to level $n + 1$. The overall structure of a computational cycle is therefore as follows:

- Solve Eqs. (9.13)–(9.15) simultaneously for ρ_μ^{n+1} and $(\rho\mathbf{u})_\sigma^{n+1}$ by means of a suitable iterative method.

- Compute ρ_σ^{n+1} by averaging the cell-centered densities ρ_μ^{n+1}.

- Compute $\mathbf{u}_\sigma^{n+1} = (\rho\mathbf{u})_\sigma^{n+1}/\rho_\sigma^{n+1}$.

- Solve Eq. (9.16) for $(\rho e)_\mu^{n+1}$.

- Compute $e_\mu^{n+1} = (\rho e)_\mu^{n+1}/\rho_\mu^{n+1}$.

- Compute $p_\mu^{n+1} = \phi(\rho_\mu^{n+1}, e_\mu^{n+1})$.

As usual, the overall stability condition for the scheme of Eqs. (9.13)–(9.16) should be reasonably well approximated by combining Eq. (4.12) with the explicit convective and diffusional stability limits derived in Sects. 7.2, 7.3, and 6.1, in which the effective net diffusivity D is taken to be the largest (or perhaps the sum) of the thermal diffusivity and the momentum diffusivities (kinematic viscosities). In one space dimension, we thereby obtain the approximate stability condition

$$\Delta t < \left(\frac{|u|}{\Delta x} + \frac{2D}{\Delta x^2} \right)^{-1} \tag{9.17}$$

which should again be further reduced by a safety factor somewhat less than unity.

The scheme of Eqs. (9.13)–(9.16) is essentially the venerable ICE ("Implicit Continuous-fluid Eulerian") method, of which numerous variants have been proposed. Depending on the precise details of how they are formulated, schemes of this type sometimes behave badly at extremely low Mach number; i.e., when the sound speed $c = (\partial p/\partial \rho)_s^{1/2}$ is very large.

There are two distinct but related reasons for this. First, equations of state of the form $p = \phi(\rho, e)$ become degenerate in the incompressible limit as $c \to \infty$, in which p is infinitely sensitive to small changes in ρ. The pressure then loses its thermodynamic significance and becomes a purely mechanical variable which can no longer be obtained from a thermodynamic equation of state. For very large but still finite values of c, p remains a legitimate thermodynamic variable in principle, but it becomes extremely sensitive to otherwise negligible variations or errors in ρ. This sensitivity is obviously conducive to erratic or unstable behavior. Conversely, however, ρ becomes extremely *insensitive* to p, so that state equations of the form $\rho = \psi(p, e)$ remain benign and well behaved in the limit $c \to \infty$, in which they simply become independent of p. The upshot is that the use of state equations of the form $p = \phi(\rho, e)$ should be avoided at very low Mach number in favor of those of the form $\rho = \psi(p, e)$. Badly behaved schemes based on the former can sometimes be algebraically transformed into well behaved schemes based on the latter.

The second source of numerical difficulties at very low Mach number is roundoff errors due to the loss of significant figures that occurs when nearly equal neighboring pressures are subtracted to compute pressure gradients. When the Mach number Ma is small, the pressure inhomogeneities in the flow field are typically of order $\delta p \sim \text{Ma}^2 \bar{p}$, where $\bar{p}$ is the uniform mean pressure level. If Ma $\sim 10^{-m}$, then even if the pressures themselves are accurately known to machine precision, the number of significant figures to which δp can be accurately represented is roughly $N - 2m$, where N is the number of significant figures retained in a floating-point number. The accuracy with which $\langle \nabla p \rangle$ can be computed is further reduced, typically by another one or two orders of magnitude, due to the fact that p normally varies by only a small fraction of δp over a single finite-difference cell. Inaccuracies of this type usually manifest themselves as noisy or irregular solutions rather than instability. According to the above estimates, they are expected to be significant only at the extremely low Mach numbers for which $m > N/2 - 1$. Consequently, this problem is rarely encountered on modern computers, but it is well to remain alert to it.

The most satisfactory way of avoiding the above problems is to reformulate Eqs. (9.10)–(9.12) in such a way that the primary computed variable is $p - \bar{p}$ rather than p itself, and to construct a numerical scheme specifically tailored to those reformulated equations. However, these are nontrivial tasks that must be approached with care. Fluid flow at very low Mach number, and *a fortiori* in the limit as $c \to \infty$, presents some physical and mathematical subtleties. The equations that result in that limit are generalizations of the usual textbook equations for incompressible flow. They no longer support sound waves, and accordingly have often been referred to as "anelastic." Several such formulations have been proposed in the literature, not all of which are well founded. However, there is little or no controversy about the basic equations of incompressible flow, the numerical solution of which will be discussed in the next section.

Equations (9.13)–(9.16) are essentially complete as they stand for theoretical purposes, but their use to compute actual numerical solutions requires the solution of the linear equation system of Eqs. (9.13)–(9.15) on every time step of the calculation. As discussed above, such systems are usually solved by iterative methods. It is absolutely essential to clearly distinguish between the implicit time-advancement scheme itself and the iteration scheme that is used to actually solve the resulting coupled system of equations for the advanced-time variables. Those two schemes are conceptually separate and distinct, and failure to recognize and clearly maintain this distinction invites confusion, errors, and inconsistencies. This danger is increased by the fact that many authors describe both those schemes together and regard them as intertwined components of a single unified algorithm. Indeed they sometimes are, since iteration schemes are sometimes specifically designed to exploit the structure of the particular coupled equations requiring solution. A further motivation for describing the time advancement and iteration schemes together is that the former is useless without the latter, so a functional self-contained computer code must contain both. (However, the code need not be entirely self-contained if it calls upon existing external modular software such as ITPACK to carry out the required iterative solution.) Unfortunately, such unified descriptions are susceptible to the type of confusion mentioned above, and they sometimes make it difficult to extract and isolate the underlying time-advancement scheme so that it can be subjected to a proper analysis. This dissection is often facilitated by the fact that many iteration schemes are time-like in nature, and can be interpreted as an artificial transient process in which the iterative solution is marched out to an artificial steady state in a pseudo-time variable. In schemes of this type, the discrete pseudo-time levels within the iteration on a given time step can be labeled by an iteration index ν analogous to the physical time index n. A useful way to see and analyze the structure of the iteration scheme, and thereby the time-advancement scheme as well, is to display the iteration index as a superscript in parentheses. Thus, for an arbitrary dependent variable f, the iterative approximation to f^{n+1} after iteration ν is denoted by $f^{(\nu)}$. The iteration scheme will involve $f^{(\nu)}$ and $f^{(\nu+1)}$ at a minimum, and possibly other pseudo-time levels as well, all of which approach f^{n+1} in the limit as $\nu \to \infty$ (provided of course that the iteration converges). Replacing $f^{(\nu)}$ and $f^{(\nu+1)}$ everywhere in the iteration scheme by f^{n+1} then reveals the time-advancement scheme that the iteration scheme is solving. Conversely, in constructing iteration schemes one may freely replace f^{n+1} by either $f^{(\nu)}$ or $f^{(\nu+1)}$ at one's discretion.

The ICE scheme of Eqs. (9.13)–(9.16) is one of the oldest and simplest schemes for low-speed flow, but various alternative schemes have been developed over the years. Another popular family of schemes for low-speed flow is based on the SIMPLE (Semi-Implicit Method for Pressure-Linked Equations) algorithm and its variants. SIMPLE schemes provide a higher degree of implicitness, which is useful in problems where the explicit diffusional stability limit is unduly restrictive and/or the explicit convective stability limit is controlled by localized regions of the flow field where $|u|/\Delta x$ is relatively large but the flow is nearly steady.

9.3. Viscous Incompressible Flow

In the absence of external forces or thermal expansion, the basic governing equations for viscous incompressible fluid flow can be written as

$$\nabla \cdot \mathbf{u} = 0 \tag{9.18}$$

$$\rho \left(\frac{\partial \mathbf{u}}{\partial t} + \mathbf{u} \cdot \nabla \mathbf{u} \right) = -\nabla p + \nabla \cdot \boldsymbol{\tau} \tag{9.19}$$

These equations are essentially a limiting special case of Eqs. (9.10)–(9.12) as $c \to \infty$, and are valid for both constant- and variable-density flows. In the latter case, the density ρ is determined by the incompressible continuity equation $\partial \rho / \partial t + \mathbf{u} \cdot \nabla \rho = 0$. However, we shall restrict attention to the most common special case, in which ρ is a given constant independent of position and time, and the viscosity coefficients are independent of temperature. Equations (9.18) and (9.19) then constitute a closed system that determines $\mathbf{u}$ and p. The energy equation Eq. (9.12) remains valid and can be used to determine the temperature T if desired, but since T does not influence the dynamics there is no need to solve for it unless it is of independent interest for some other reason.

A conspicuous feature of Eqs. (9.18) and (9.19) is that p is not determined by a time evolution equation, but rather is implicitly determined by Eq. (9.18), which plays the role of a constraint that the velocity field must satisfy. The pressure instantaneously adapts to the evolving velocity field in such a way as to satisfy that constraint. This is reflected in the fact that p satisfies a Poisson equation, which can be derived by taking the divergence of Eq. (9.19) and combining the result with Eq. (9.18). Equation (9.18) can then be replaced by the Poisson equation for p, the solution of which can in principle be substituted back into Eq. (9.19) to obtain an evolution equation for the velocity field alone. Alternatively, the pressure can be immediately eliminated by taking the curl of Eq. (9.19), which leads to the vorticity-stream function formulation of incompressible flow. Formal manipulations of this type are useful for various theoretical purposes, but experience has shown that they are rarely advantageous for computational purposes. In most situations, it is preferable to simply approximate and solve Eqs. (9.18) and (9.19) directly for the so-called *primitive variables* $\mathbf{u}$ and p.

Once again we seek a scheme with the minimum implicitness required for stability. The essential features of such a scheme can be inferred from, and indeed are almost uniquely determined by, some very basic considerations. As discussed in the previous section, Eqs. (9.18) and (9.19) no longer support sound waves due to the fact that the pressure in Eq. (9.19) is now a purely mechanical variable which is no longer related to the density by a thermodynamic equation of state. Moreover, Eqs. (9.18) and (9.19) do not involve $\partial p / \partial t$, so the pressure cannot be advanced in time by means of its time derivative. In fact, the only place p appears in the equations at all is in the pressure gradient term in Eq. (9.19). If this term were differenced in a purely explicit manner, the scheme would not involve p^{n+1} in any way, so there would be no way to compute it. In order to advance the pressure in time, the temporal difference approximation to ∇p in Eq. (9.19) must be at least partially implicit so that p^{n+1} appears in the difference scheme. This in turn implies that the temporal difference approximation to $\nabla \cdot \mathbf{u}$ in Eq. (9.18) must likewise be at least partially implicit, since the difference approximation to Eq. (9.19) alone does not provide enough equations to determine both $\mathbf{u}^{n+1}$ and p^{n+1}. But once $\nabla \cdot \mathbf{u}$ is partially implicit, it might as well be fully implicit, which entails no additional labor and has the significant advantage of satisfying Eq. (9.18) exactly at each time level of the calculation. We therefore difference Eq. (9.18) in a fully implicit manner to obtain

$$\langle \nabla \cdot \mathbf{u} \rangle_\mu^{n+1} = 0 \tag{9.20}$$

where the spatial differencing symbolized by $\langle \cdots \rangle$ is performed by the finite-volume method as before.

Except for the pressure gradient, the remaining terms in Eq. (9.19) can evidently be differenced in a fully explicit manner, with the result

$$\frac{\mathbf{u}_\sigma^{n+1} - \mathbf{u}_\sigma^n}{\Delta t} + \langle \mathbf{u} \cdot \nabla \mathbf{u} \rangle_\sigma^n = -\frac{1}{\rho}\langle \nabla p^\star \rangle_\sigma + \frac{1}{\rho}\langle \nabla \cdot \boldsymbol{\tau} \rangle_\sigma^n \qquad (9.21)$$

where $p^\star = \gamma p^{n+1} + (1 - \gamma)p^n$ as discussed above, with $0 < \gamma \leq 1$, and the spatial differencing is again performed by the finite-volume method. If desired, the term $\mathbf{u} \cdot \nabla \mathbf{u}$ can be rewritten as $\nabla \cdot (\mathbf{uu})$ so that it can be differenced in a conservative manner as discussed in Sect. 7.5. In either case, it will again require upwinding for stability.

We now observe that once the spatial differencing has been specified in detail, Eqs. (9.20) and (9.21) constitute a closed system of equations that uniquely determine the values of $\mathbf{u}^{n+1}$ and $p^\star$, regardless of the value of the as yet unspecified positive weighting factor γ. It follows that the velocity field is entirely indifferent to the value of γ; i.e., to the precise degree of implicitness with which the pressure gradient is differenced. Varying γ will therefore produce small changes of order Δt in the time history of the pressure field, but this will have no effect whatever on the velocity field. As a result, the precise value of γ is essentially immaterial, so we may as well adopt the simplest allowed choice $\gamma = 1$, thereby setting $p^\star = p^{n+1}$ and hence approximating the pressure gradient in a fully implicit manner. Equation (9.21) then becomes

$$\frac{\mathbf{u}_\sigma^{n+1} - \mathbf{u}_\sigma^n}{\Delta t} + \langle \mathbf{u} \cdot \nabla \mathbf{u} \rangle_\sigma^n = -\frac{1}{\rho}\langle \nabla p \rangle_\sigma^{n+1} + \frac{1}{\rho}\langle \nabla \cdot \boldsymbol{\tau} \rangle_\sigma^n \qquad (9.22)$$

Equations (9.20) and (9.22) then define a basic partially implicit time advancement scheme for viscous incompressible flow. This scheme is the heart of the MAC ("Marker and Cell") method. The nomenclature is now of historical interest only, but the scheme itself is timeless. Equations (9.20) and (9.22) constitute a closed coupled linear system of equations for the advanced-time variables $\mathbf{u}_\sigma^{n+1}$ and p_μ^{n+1}. Those equations can be algebraically combined in such a way as to eliminate $\mathbf{u}_\sigma^{n+1}$, which results in a discrete Poisson equation for the pressures p_μ^{n+1} alone. However, experience has again shown that this is not particularly advantageous, and indeed introduces unnecessary complications that make it more difficult to impose boundary conditions on the fluid velocity.

The implicitness in Eqs. (9.20) and (9.22) bears an obvious similarity to that of Eqs. (9.13) and (9.14), and shows that the ICE method is the natural generalization of the MAC method to low-speed flow. As shown in the previous section, however, the ICE method can be independently motivated, so that the MAC method can equally well be regarded as the incompressible limit of the ICE method. Since the explicit parts of the two schemes are essentially the same, their stability conditions are also expected to be the same, so that Eq. (9.17) is an equally valid estimate for the stability limit of Eqs. (9.20) and (9.22).

As usual, Eqs. (9.20) and (9.22) are ordinarily solved by iterative methods, of which the traditional historical choices were essentially equivalent to SOR. The most popular of those choices has an intuitively appealing physical interpretation, the essence of which is as follows. In accordance with the notational convention described in the previous section, the intermediate approximations to $\mathbf{u}_\sigma^{n+1}$ and p_μ^{n+1} after iteration ν will be denoted by $\mathbf{u}_\sigma^{(\nu)}$ and $p_\mu^{(\nu)}$, respectively. During the iteration (i.e., prior to its convergence), those approximations do not of course satisfy Eqs. (9.20) and (9.22). Now if the intermediate value of $\langle \nabla \cdot \mathbf{u} \rangle_\mu^{(\nu)}$ is positive in a particular cell μ, this implies that the fluid is unphysically expanding, which in turn suggests that p_μ in that cell should be reduced in order to pull more fluid into it. Conversely, if $\langle \nabla \cdot \mathbf{u} \rangle_\mu^{(\nu)} < 0$ then the fluid is unphysically compressing, in which case p_μ should be increased to push fluid out of it. These desired effects can be achieved by setting

$$\delta p_\mu \equiv p_\mu^{(\nu+1)} - p_\mu^{(\nu)} = -\Omega \langle \nabla \cdot \mathbf{u} \rangle_\mu^{(\nu)} \tag{9.23}$$

where the positive coefficient Ω is constrained by the requirement that the iteration scheme remain stable; i.e., it must not diverge. The values to which Ω is thereby restricted can be determined by means of a Fourier stability analysis in which n is replaced by ν.

Once the values of $p_\mu^{(\nu+1)}$ have been computed via Eq. (9.23), the corresponding values of $\mathbf{u}_\sigma^{(\nu+1)}$ are then determined by

$$\frac{\mathbf{u}_\sigma^{(\nu+1)} - \mathbf{u}_\sigma^n}{\Delta t} + \langle \mathbf{u} \cdot \nabla \mathbf{u} \rangle_\sigma^n = -\frac{1}{\rho} \langle \nabla p \rangle_\sigma^{(\nu+1)} + \frac{1}{\rho} \langle \nabla \cdot \boldsymbol{\tau} \rangle_\sigma^n \tag{9.24}$$

or equivalently

$$\delta \mathbf{u}_\sigma \equiv \mathbf{u}_\sigma^{(\nu+1)} - \mathbf{u}_\sigma^{(\nu)} = -\frac{\Delta t}{\rho} \langle \nabla(\delta p) \rangle_\sigma \tag{9.25}$$

It is obvious by inspection that if the iteration scheme of Eqs. (9.23) and (9.24) is convergent (i.e., if $u_\sigma^{(\nu)}$ and $p_\mu^{(\nu)}$ approach finite limiting values as $\nu \to \infty$), then those values satisfy Eqs. (9.20) and (9.22) and therefore determine u_σ^{n+1} and p_μ^{n+1}.

The arithmetic calculations required by Eqs. (9.23) and (9.25) must be performed on every cell μ and cell face σ in the computing mesh. On serial processors, those calculations are of course performed sequentially rather than simultaneously. This is ordinarily done by systematically sweeping or looping through the mesh and proceeding from one cell to the next. In calculations of this type, the convergence rate of the iteration scheme can be significantly accelerated by immediately incrementing the velocities on all of the faces of cell μ to include the effects of δp_μ as soon as the latter has been computed. The corresponding velocity increments can readily be inferred from Eq. (9.25) once the spatial difference approximation to the pressure gradient has been specified. On cell faces where they are available, the resulting partially updated velocities are then used in Eq. (9.23), in lieu of $u_\sigma^{(\nu)}$, to evaluate $\langle \nabla \cdot u \rangle$ and δp on the next cell. These modifications convert what was essentially a weighted Jacobi iteration scheme into SOR, which converges much more rapidly for optimal or nearly optimal values of Ω. It also has the secondary advantage of requiring less storage, since after p and u have been fully or even partially updated on the current iteration, their values on the previous iteration are no longer required and hence need not be saved. This immediate use of all the available new information as soon as it has been computed is a standard simple technique for accelerating convergence, and is an essential feature of the Gauss–Seidel and SOR methods. It is also useful in accelerating the rate at which time-marching transient calculations relax to a steady-state solution, as discussed in Sect. 1.1.

BIBLIOGRAPHY

Abbott, M. B. and D. R. Basco, *Computational Fluid Dynamics: An Introduction for Engineers* (Wiley/Longman, New York/Harlow, 1989).

Ames, W. F., *Numerical Methods for Partial Differential Equations*, 2nd ed. (Academic, New York, 1977).

Anderson, D. A., J. C. Tannehill, and R. H. Pletcher, *Computational Fluid Mechanics and Heat Transfer* (Hemisphere, New York, 1984).

Anderson, E., Z. Bai, C. Bischof, S. Blackford, J. Demmel, J. Dongarra, J. Du Croz, A. Greenbaum, S. Hammarling, A. McKenney, and D. Sorensen, *LAPACK Users' Guide*, 3rd ed. (SIAM, Philadelphia, 1999).

Anderson, J. D., Jr., *Computational Fluid Dynamics* (McGraw-Hill, New York, 1995).

Axelsson, O., *Iterative Solution Methods* (Cambridge U. P., Cambridge, 1994).

Baines, M. J., *Moving Finite Elements* (Oxford U. P., Oxford & New York, 1994).

Baker, A. J., *Finite Element Computational Fluid Mechanics* (McGraw-Hill, New York, 1983).

Barker, V. A., L. S. Blackford, J. Dongarra, J. Du Croz, S. Hammarling, M. Marinova, J. Waśniewski, and P. Yalamov, *LAPACK95 Users' Guide* (SIAM, Philadelphia, 2001).

Barrett, R., M. Berry, T. F. Chan, J. Demmel, J. Donato, J. Dongarra, V. Eijkhout, R. Pozo, C. Romine, and H. van der Vorst, *Templates for the Solution of Linear Systems: Building Blocks for Iterative Methods* (SIAM, Philadelphia, 1994).

Birkhoff, G., *The Numerical Solution of Elliptic Equations* (SIAM, Philadelphia, 1972).

Birkhoff, G. and R. E. Lynch, *Numerical Solution of Elliptic Problems* (SIAM, Philadelphia, 1984).

Blazek, J., *Computational Fluid Dynamics: Principles and Applications* (Elsevier, Amsterdam & Oxford, 2001).

Bose, T. K., *Computational Fluid Dynamics* (Wiley, New York, 1988).

Briggs, W. L., *A Multigrid Tutorial* (SIAM, Philadelphia, 1987).

Burgers, J. M., *Flow Equations for Composite Gases* (Academic, New York, 1969).

Canuto, C., M. Y. Hussaini, A. Quarteroni, and T. A. Zang, *Spectral Methods in Fluid Mechanics* (Springer, Berlin, 1987).

Cebeci, T., J. P. Shao, F. Kafyeke, and E. Laurendeau, *Computational Fluid Dynamics for Engineers: From Panel to Navier–Stokes Methods with Computer Programs* (Springer/Horizons, Berlin/Long Beach, 2005).

Chung, T. J., *Computational Fluid Dynamics* (Cambridge U. P., Cambridge, 2002).

Coleman, T. F. and C. Van Loan, *Handbook for Matrix Computations* (SIAM, Philadelphia, 1988).

Date, A. W., *Introduction to Computational Fluid Dynamics* (Cambridge U. P., Cambridge, 2005).

Duff, I. S., A. M. Erisman, and J. K. Reid, *Direct Methods for Sparse Matrices* (Oxford U. P., New York, 1986).

Faddeev, D. K. and V. N. Faddeeva, *Computational Methods of Linear Algebra* (Freeman, San Francisco & London, 1963).

Faddeeva, V. N., *Computational Methods of Linear Algebra* (Dover, New York, 1959).

Feistauer, M., J. Felcman, and I. Straškraba, *Mathematical and Computational Methods for Compressible Flow* (Oxford U. P., Oxford, 2003).

Ferziger, J. H. and M. Perić, *Computational Methods for Fluid Dynamics*, 3rd ed. (Springer, Berlin, 2002).

Finlayson, B. A., *The Method of Weighted Residuals and Variational Principles* (Academic, New York, 1972).

Fletcher, C. A. J., *Computational Techniques for Fluid Dynamics*, 2nd ed., Vols. I and II (Springer, Berlin, 1991).

Forsythe, G. E. and C. B. Moler, *Computer Solution of Linear Algebraic Systems* (Prentice Hall, Englewood Cliffs, NJ, 1967).

Forsythe, G. E. and W. R. Wasow, *Finite-Difference Methods for Partial Differential Equations* (Wiley, New York, 1960).

Garg, V. K. ed., *Applied Computational Fluid Dynamics* (Dekker, New York, 1998).

Gear, C. W., *Numerical Initial Value Problems in Ordinary Differential Equations* (Prentice Hall, Englewood Cliffs, NJ, 1971).

Godunov, S. K. and V. S. Ryabenki, *Theory of Difference Schemes: An Introduction* (North-Holland, Amsterdam, 1964).

Golub, G. H. and C. F. Van Loan, *Matrix Computations* (Johns Hopkins U. P., Baltimore, 1983).

Gottlieb, D. and S. A. Orszag, *Numerical Analysis of Spectral Methods: Theory and Applications* (SIAM, Philadelphia, 1977).

Gresho, P. M. and R. L. Sani, *Incompressible Flow and the Finite Element Method*, Vols. 1 and 2 (Wiley, West Sussex, 2000).

Griebel, M., T. Dornseiter, and T. Neunhoeffer, *Numerical Simulation in Fluid Dynamics: A Practical Introduction* (SIAM, Philadelphia, 1998).

Hageman, L. A. and D. M. Young, *Applied Iterative Methods* (Academic, New York, 1981).

Hirsch, C., *Numerical Computation of Internal and External Flows: Fundamentals of Computational Fluid Dynamics*, 2nd ed. (Elsevier, Amsterdam, 2007).

Hoffmann, K. A. and S. T. Chiang, *Computational Fluid Dynamics*, Vol. 1, 4th ed. (Engineering Education System, Wichita, KS, 2000).

Holt, M., *Numerical Methods in Fluid Dynamics*, 2nd ed. (Springer, Berlin, 1984).

Hornbeck, R. W., *Numerical Marching Techniques in Fluid Flows with Heat Transfer* (NASA SP-297, Washington, DC, 1973).

Jou, D., J. Casas-Vazquez, and G. Lebon, *Extended Irreversible Thermodynamics* (Springer, Berlin, 2001).

Kotake, S. and K. Hijikata, *Numerical Simulations of Heat Transfer and Fluid Flow on a Personal Computer* (Elsevier, Amsterdam, 1993).

Laney, C. B., *Computational Gasdynamics* (Cambridge U. P., Cambridge, 1998).

Lapidus, L. and G. F. Pinder, *Numerical Solution of Partial Differential Equations in Science and Engineering* (Wiley, New York, 1999).

LeVeque, R. J., *Finite Volume Methods for Hyperbolic Problems* (Cambridge U. P., London, 2002).

Lohner, R., *Applied Computational Fluid Dynamics Techniques* (Wiley, Chichester, 2008).

Lomax, H., T. H. Pulliam, and D. W. Zingg, *Fundamentals of Computational Fluid Dynamics* (Springer, Berlin, 2001).

Marchuk, G. I. and V. V. Shaidurov, *Difference Methods and Their Extrapolations* (Springer, New York, 1983).

Milne, W. E., *Numerical Solution of Differential Equations*, 2nd ed. (Dover, New York, 1970).

Morton, K. W. and D. F. Mayers, *Numerical Solution of Partial Differential Equations* (Cambridge U. P., Cambridge, 1994).

Oran, E. S. and J. P. Boris, *Numerical Simulation of Reactive Flow*, 2nd ed. (Cambridge U. P., Cambridge, 2001).

Ortega, J. M. and W. C. Rheinboldt, *Iterative Solution of Nonlinear Equations in Several Variables* (Academic, New York, 1970).

Patankar, S. V., *Numerical Heat Transfer and Fluid Flow* (Hemisphere, Washington, 1980).

Patankar, S. V. and D. B. Spalding, *Heat and Mass Transfer in Boundary Layers: A general calculation procedure*, 2nd ed. (Intertext, London, 1970).

Peyret, R. ed., *Handbook of Computational Fluid Mechanics* (Academic, San Diego/London, 1996).

Peyret, R. and T. D. Taylor, *Computational Methods for Fluid Flow* (Springer, New York, 1983).

Pozrikidis, C., *Introduction to Theoretical and Computational Fluid Dynamics* (Oxford U. P., New York, 1997).

Rice, J. R., *Matrix Computations and Mathematical Software* (McGraw-Hill, New York, 1981).

Richtmyer, R. D. and K. W. Morton, *Difference Methods for Initial-Value Problems*, 2nd ed. (Wiley/Interscience, New York, 1967).

Roache, P. J., *Computational Fluid Dynamics* (Hermosa, Albuquerque, 1985).

Roache, P. J., *Fundamentals of Computational Fluid Dynamics* (Hermosa, Albuquerque, 1998).

Saad, Y., *Iterative Methods for Sparse Linear Systems*, 2nd ed. (SIAM, Philadelphia, 2003).

Saul'yev, V. K., *Integration of Equations of Parabolic Type by the Method of Nets* (Pergamon/Macmillan, New York, 1964).

Schultz, M. ed., *Elliptic Problem Solvers* (Academic, New York, 1981).

Shampine, L. F. and M. K. Gordon, *Computer Solution of Ordinary Differential Equations: The Initial Value Problem* (Freeman, San Francisco, 1975).

Shashkov, M., *Conservative Finite-Difference Methods on General Grids* (CRC Press, Boca Raton, FL, 1996).

Shaw, C. T., *Using Computational Fluid Dynamics* (Prentice Hall, Hemel Hempstead, 1992).

Shokin, Y. I., *The Method of Differential Approximation* (Springer, Berlin, 1983).

Shyy, W., H. S. Udaykumar, M. M. Rao, and R. W. Smith, *Computational Fluid Dynamics with Moving Boundaries* (Dover, New York, 2007).

Smith, G. D., *Numerical Solution of Partial Differential Equations: Finite Difference Methods*, 3rd ed. (Oxford U. P., Oxford, 1985).

Stewart, G. W., *Introduction to Matrix Computations* (Academic, New York, 1973).

Tannehill, J. C., D. A. Anderson, and R. H. Pletcher, *Computational Fluid Mechanics and Heat Transfer*, 2nd ed. (Taylor & Francis, Philadelphia/ London, 1997).

Traub, J. F., *Iterative Methods for the Solution of Equations* (Chelsea, New York, 1982).

Tu, J., G. H. Yeoh, and C. Liu, *Computational Fluid Dynamics: A Practical Approach* (Butterworth-Heinemann, Amsterdam/Boston, 2008).

van der Vorst, H. A., *Iterative Krylov Methods for Large Linear Systems* (Cambridge U. P., Cambridge, 2003).

Varga, R., *Matrix Iterative Analysis* (Prentice Hall, Englewood Cliffs, NJ, 1962).

Versteeg, H. K. and W. Malalasekera, *An Introduction to Computational Fluid Dynamics: The Finite Volume Method*, 2nd ed. (Pearson/Prentice Hall, Harlow/New York, 2007).

Vichnevetsky, R. and J. B. Bowles, *Fourier Analysis of Numerical Approximations of Hyperbolic Equations* (SIAM, Philadelphia, 1982).

Warsi, Z. U. A., *Fluid Dynamics: Theoretical and Computational Approaches*, 2nd ed. (CRC Press, Boca Raton, FL, 1999).

Weaver, H. J., *Applications of Discrete and Continuous Fourier Analysis* (Wiley-Interscience, New York, 1983).

Weaver, H. J., *Theory of Discrete and Continuous Fourier Analysis* (Wiley-Interscience, New York, 1989).

Wendt, J. F. ed., *Computational Fluid Dynamics: An Introduction*, 3rd ed. (Springer, Berlin, 2009).

Wesseling, P., *Principles of Computational Fluid Dynamics* (Springer, Berlin, 2001).

Yanenko, N. N., *The Method of Fractional Steps* (Springer, New York, 1971).

Young, D. M., *Iterative Solution of Large Linear Systems* (Dover, New York, 2003).

Zikanov, O., *Essential Computational Fluid Dynamics* (Wiley, Hoboken, NJ, 2010).

INDEX

ADI methods, 34, 67
  stability, 68, 69
advection, 71
ALE methods, 104
alternate-direction methods, 34, 67
amplification factor, 43
  complex, 73
anelastic equations, 114

backward differencing, 14
BTCS schemes, 62
  convection, 74
  diffusion, 65
Burgers equation, 41

cell, 12
  corner, 12
  edge, 12
  face, 12
  momentum, 110
  vertex, 12
centered differencing, 15
CFL condition, 72
  acoustic, 95, 96, 109
  convective, 96, 109
chaos, 46
characteristics, 72
checkerboarding, 13, 15, 95
chemical reactions, 55, 57
Cloutman, L. D., 107
compressible flow, 26, 104, 108
compression, 83
conjugate gradient methods, 35
conjugate residual methods, 35

conservation, local, 85
conservative form, 82
conservative methods, 83, 84
consistency, 18
continuity equation, 83
convection, 71
convective derivative, 92
convergence, 26
  iteration, 115, 119
  rate, 26, 119
Courant condition, 72
Courant number, 72
  acoustic, 94
  convective, 77
Cramer's rule, 57
Crank–Nicholson method
  diffusion, 62, 65
  oscillations, 53, 66, 69
  source terms, 52
CTCS schemes, 62
cycle, 12

debugging, 27
diffusion equation, 61
diffusivity, 61
  artificial, 75, 77–79, 108
  negative, 75
discretization error, 15
dispersion, 80
dispersion errors, 79, 80, 101
dispersion relation, 81
  diffusional, 64, 75
divergence form, 82
divergence theorem, 85, 88